Bulldozer Capitalism

DISLOCATIONS

General Editors: August Carbonella, *Memorial University of Newfoundland;* Don Kalb, *University of Bergen & Utrecht University;* Linda Green, *University of Arizona*

The immense dislocations and suffering caused by neoliberal globalization, the retreat of the welfare state in the last decades of the twentieth century, and the heightened military imperialism at the turn of the twenty-first century have raised urgent questions about the temporal and spatial dimensions of power. Through stimulating critical perspectives and new and cross-disciplinary frameworks that reflect recent innovations in the social and human sciences, this series provides a forum for politically engaged and theoretically imaginative responses to these important issues of late modernity.

Recent volumes:

Volume 31
Bulldozer Capitalism: Accumulation, Ruination, and Dispossession in Northeastern Turkey
Erdem Evren

Volume 30
Facing the Crisis: Ethnographies of Work in Italian Industrial Capitalism
Edited by Fulvia D'Aloisio and Simone Ghezzi

Volume 29
Big Capital in an Unequal World: The Micropolitics of Wealth in Pakistan
Rosita Armytage

Volume 28
Fifty Years of Peasant Wars in Latin America
Edited by Leigh Binford, Lesley Gill and Steve Striffler

Volume 27
Brazilian Steel Town: Machines, Land, Money and Commoning in the Making of the Working Class
Massimiliano Mollona

Volume 26
Claiming Homes: Confronting Domicide in Rural China
Charlotte Bruckermann

Volume 25
Democracy Struggles: NGOs and the Politics of Aid in Serbia
Theodora Vetta

Volume 24
Worldwide Mobilizations: Class Struggles and Urban Commoning
Edited by Don Kalb and Massimiliano Mollona

Volume 23
The Revolt of the Provinces: Anti-Gypsyism and Right-Wing Politics in Hungary
Kristóf Szombati

Volume 22
Frontiers of Civil Society: Government and Hegemony in Serbia
Marek Mikuš

For a full volume listing, please see the series page on our website: https://www.berghahnbooks.com/series/dislocations

Bulldozer Capitalism

Accumulation, Ruination, and Dispossession in Northeastern Turkey

Erdem Evren

berghahn
NEW YORK • OXFORD
www.berghahnbooks.com

First published in 2022 by
Berghahn Books
www.berghahnbooks.com

Library of Congress Cataloging-in-Publication Data
A C.I.P. cataloging record is available from the Library of Congress
Library of Congress Cataloging in Publication Control Number: 2022006151

British Library Cataloguing in Publication Data
A catalogue record for this book is available from the British Library

ISBN 978-1-80073-280-3 hardback
ISBN 978-1-80073-474-6 ebook

https://doi.org/10.3167/9781800732803

CONTENTS

ILLUSTRATIONS

Figures

Maps

Acknowledgments

Bulldozer Capitalism would not have been possible without the support and encouragement of several friends, colleagues, interlocutors, and family members. Even though I am solely responsible for the book's shortcomings, I shall forever remain indebted to them. I would like to thank first and foremost those who allowed me to conduct research in Yusufeli and tolerated my awkward presence in their lives: Zeycan, Hera, Osman, Ali, Serdar, Ayhan K., Onur, Aysel, Recep, Ferhan, and Ayhan. They have enriched my life in so many unexpected ways.

At the Leibniz-Zentrum Moderner Orient (ZMO), the institutional home of this project for several years, Ulrike Freitag, Katharina Lange, Silke Nagel, Ali Nobil Ahmad, Jeanine Dağyeli, Yasmine Berriane, Patrick Schukalla, and Paolo Giabazzi, among many others, made this book possible and knowingly or unknowingly helped me to sharpen its arguments. Silvia Schröcker, Sagi Rotfogel, Teo Schlögl, and Sophie Wilske have been excellent student assistants. Funds for this project were provided by the Federal Ministry of Education and Research, Germany (funding code 01UG0713). At the University of Toronto where I spent an academic year with the support of a Volkswagen Foundation fellowship, Tania Li, Andrea Mühlebach, Shozab Raza, Hadia Akhtar Khan, Stephen Campbell, Lukas Ley, Sardar Saadi, and Leyla Safta-Zecheria taught me more than I could ever ask for. I would also like to thank Todd Sanders and the members of his dissertation writing group for their help when needed. My students at the University of Toronto and Humboldt University, Berlin also deserve credit for their willingness to engage with some of the ideas that I explore in this book.

Many other scholars have supported my work over the years by inviting me to take part in workshops, edited volumes, and reading groups. Some of them became close collaborators, intellectual interlocutors, and dear friends. I wish to express my gratitude in particular to Alice von Bieberstein, Sinan Erensü, Cemil Aksu, Özge Yaka, Özlem Biner, and Yael Navaro. I also wish to thank Enis Oktay, Emre Gözgü, Theo Barry-Born, André Bank, Danilo Caputo, members of the

Ishtirak group, Elske Rosenfeld, Mira Wallis, May Parlar, and Alexandra von der Linde for their friendship and guidance. My parents, Nil Erencan and Akın Evren, and my brother, Erencan Evren, have been waiting for this book for quite a long time. I know it will make them proud.

Earlier versions of Chapter 1 and Chapter 5 were published by the *Journal of Water History* and the University of Pennsylvania Press, respectively. Harry Eagles, Tom Bonnington, Anthony Mason, Elizabeth Martinez and Keara Hagerty from Berghahn have been unfailing in their support. Alice von Bieberstein, Sinan Erensü, Özge Yaka, and two anonymous reviewers have read drafts and pushed me to clarify my arguments.

I am grateful beyond words to Nadine Püschel and Milo Evren who continuously endured my mood swings, long absences, and periods of indecision during the writing of this book. Without their love and patience, this book, which I dedicate to them, would never have seen the light. Thank you to both of you!

Abbreviations

AKP	Justice and Development Party
BBP	Grand Unity Party
CBT	Central Bank of Turkey
CHP	Republican People's Party
CUP	Committee for the Union and Progress
DPT	Department of State Planning
DSI	General Directorate of the State Hydraulic Works
ECA	Export Credit Agency
EIA	Environmental Impact Assessment
HDP	Peoples' Democratic Party
HEPP	Hydroelectric Power Plant
IMF	International Monetary Fund
IFC	International Finance Corporations
İP	Good Party
IŞKUR	Turkish Employment Agency
MHP	Nationalist Action Party
PKK	Kurdistan Workers' Party
TOKI	Mass Housing and Public Administration
WCD	World Commission on Dams

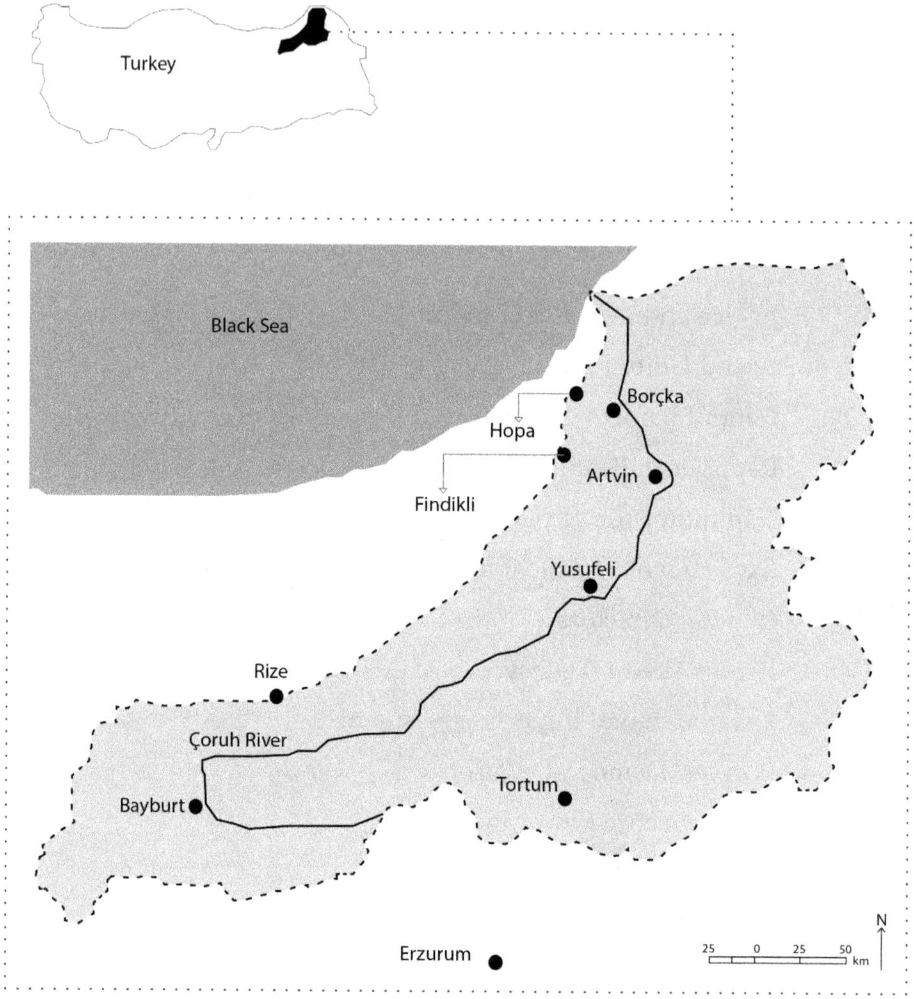

Map 0.1. Map of the eastern Black Sea region of Turkey. © Erdem Evren.

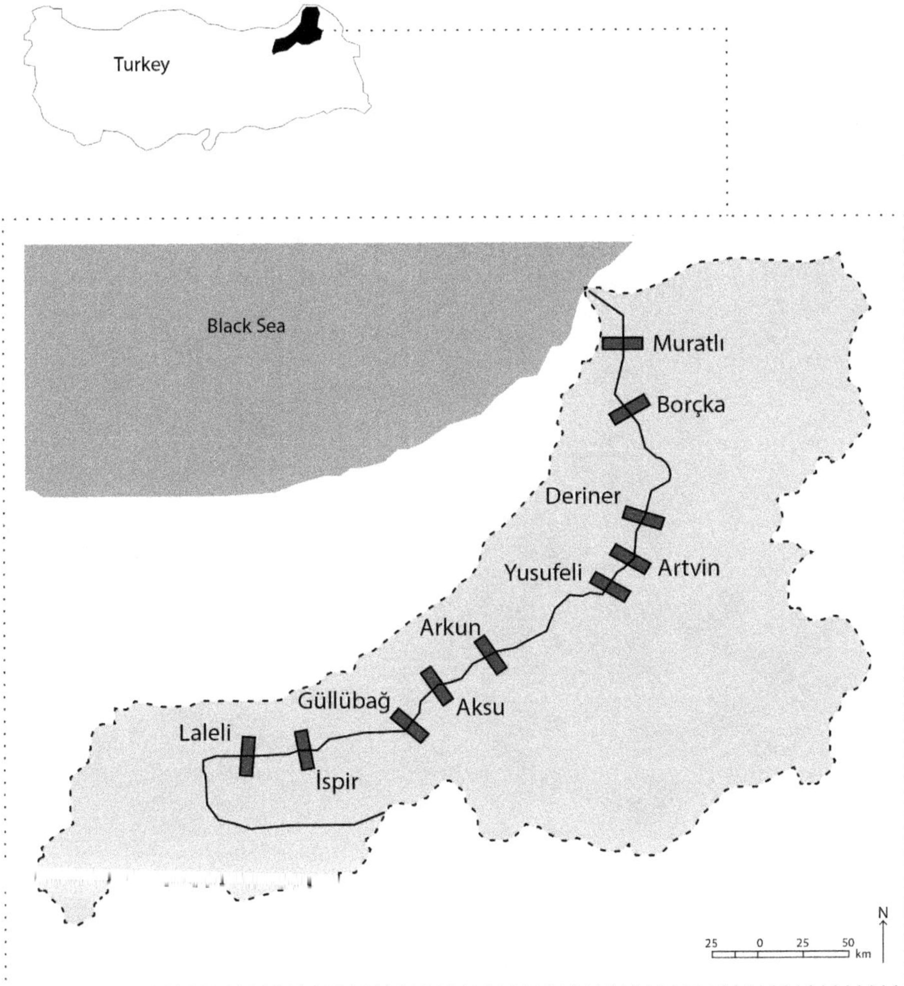

Map 0.2. Map of large dam projects on the Çoruh River. © Erdem Evren.

INTRODUCTION

> In the dark times,
> will there also be singing?
> Yes, there will be singing
> about the dark times.
> —Bertolt Brecht, "Motto"

Politics in a Resource Frontier, Interrupted

Starting from the end of the 2000s, Turkey's countryside began to grab the headlines with reports on protests against resource extraction and infrastructural development projects. The planning and construction of hydroelectric, thermal, geothermal, and coal-burning plants, electricity transmission lines and mining facilities were being met with fierce opposition by rural communities across the country. Protest actions where men and women, old and young, and farmers and pensioners together clashed with the police and gendarmerie for hours were not rare. Occasionally, company meetings were disrupted, and construction vehicles were damaged or set on fire. Soon, they were joined by leftwing activists, journalists, lawyers, artists, and academics from urban centers who were at once intrigued and inspired by the intensity and prevalence of these struggles. This time they were ready to invest their political hopes on "peasants."

I jumped on this bandwagon at the end of 2011 when I traveled for two weeks in the eastern Black Sea region (see Map 0.1). This part of Turkey had by then already become a hotbed of activism especially against the construction of small-scale run-of-the-river-type hydroelectric power plants (HEPPs).[1] In the coastal towns, I spoke to one village community after another about the court cases they had brought against private companies who were preparing to build tens of HEPPs on virtually every single river and rivulet crossing their valleys. Socialist activists in Hopa and Fındıklı explained that the state had leased the use rights of river sections to these corporations for energy production for forty-nine years, thus setting in motion the commodification and loss of water, and the destruction of land. In the regional capital Artvin, I witnessed residents reviving their successful fight of the previous decade against the renewed plans for gold mining in a nearby hilltop called Cerrattepe. A new company, this time a domestic one strongly backed by the ruling Justice and Development Party (AKP), was once again pushing for the opening of pits against residents' strong opposition that would eventually snowball into a city-wide confrontation with the gendarmerie.[2]

Located further south, about one and a half hours away from Artvin by bus (provided that the road is not closed or did not collapse), the political mood in and around the town of Yusufeli, by contrast, was congruent with its fate, which was by then more or less sealed: submergence of the entire town and its seventeen villages, displacement of thousands of people, and the destruction of nearly all agricultural land with the coming construction of the Yusufeli Dam[3]—the tallest of ten large dams planned or already completed on the Çoruh River as part of the Çoruh Energy Plan (see Map 0.2). My earliest interlocutors, too, were of an antipodean ideological disposition, veering toward ultra-nationalism and conservatism. I was struck, if not startled, by their politics, the scale of destruction awaiting Yusufeli and its surroundings, and how little attention all this was receiving among the activists from the region, let alone the general public.

Not long after, I would be astounded to find out as well that roughly for a decade Yusufeli's residents had in fact succeeded in preventing the start of the project. When the beginning of construction by an international consortium was unceremoniously announced in 1997, they quickly mobilized around a cultural association to launch a powerful anti-displacement campaign.[4] On the local level, demonstrations were held to protest against state agencies and regional governors, and court cases were brought against various ministries and administrative bodies. On the international level, there was a concerted effort in collaboration with European NGOs to dissuade export credit agencies (ECAs) from providing loans to construction companies involved in the project. To this end, activists utilized international standards and guidelines on resettlement and scandalized the absence of an environmental impact assessment (EIA) report. As a result, the realization of the project was deferred as one international consortium was compelled to withdraw after another.

After the Turkish state officially announced in 2010 that it would finance the project exclusively however, the earlier activist energy in the town fizzled out to be eventually replaced by negotiating with the state as the dominant form of collective action. Locals intermittently continued in the following years to resort to legal means to oppose the HEPPs planned in their villages, as in other parts of the region. The cultural association, which had by then been reduced to a handful of activists, even continued to try bringing new court cases to the Constitutional Court of Turkey and the European Court of Human Rights for the cancellation of the Yusufeli Dam project. The primary concern for the majority of residents however has become the tireless bargaining over the worth of: a two-story house, a fruit orchard, the business relations of a shop owner who has spent his entire life in the town center, or the loyalty to the state of an entire community planned to be sacrificed for energy production.

This book then is about politics in a resource frontier, on its formations and blockage at the intersections of infrastructure, economy, and identity. By politics, I understand, following Antonio Gramsci (1971), first and foremost the capacity to engage in critical thought and action in the practices of everyday life. Embedded in the common sense in incoherent and contradictory ways, this embryonic form of critique could become historically effective, according to him, only if it is rendered visible and societal forces are connected with one another. Gramsci's anthropologist readers (Smith 1999; Crehan 2002; Li 2007, 2019) have argued, and in different ways demonstrated, that an ethnography of politics necessarily requires both the study of this work of organization and articulation, and the conditions for its interruption. Following them, my first aim in this book is to examine how everyday forms of discontent come to be contained and rendered ineffective even after they consistently become the content for an effective political intervention. I am interested in understanding why this semirural community gradually began to give consent and take part in a project that would result in its economic dispossession and involuntary displacement. To this end, I explore the circumstances and forces that first enabled and then hindered its members' capacity to forge connections and to make explicit their critical challenge against the infrastructure's injuries and injustices.

By resource frontier, I refer not only to those sites where the state and private companies carry out or facilitate the appropriation of natural resources but more broadly make a gesture towards an entire web where the logic of assigning monetary values to things and relations proliferate and seep into other spheres of life under capitalist development.[5] My ethnographic research has shown that dam constructions along the Çoruh River fueled the residents' aspirations to pursue different kinds of value formation including but not limited to employment, conspicuous consumption, and speculation. Later, I will address in more detail these practices and imaginaries that exceed yet emerge in relation to the state's and capital's extraction of value from land and water. By invoking them here, I simply intend to counter a narrative inspired by James Scott's seminal work *Seeing Like a State* (1998) that still continues to prevail the thinking behind much of the literature on large infrastructural projects: grand schemas of improvement sever local people from everyday forms of knowing and doing that have been central to their survival for centuries. Contrary to those accounts that merely pit high modernist schemas such as mega dams and their promises of progress against the figure of "surplus people" or "uninhabitants" often marked by self-contained or Indigenous ways of living, my second aim in the book is to show that capitalist development instead works on and through the desires of its target communities (Li 2005).

Through what mediums and linkages do people's aspirations encounter the logics and operations of the state and capital? And what does subscribing to capitalist development actually do to its target communities? This book makes two related claims. First, I argue that destruction itself becomes the conduit for realizing economic and political desires when the infrastructural technologies deployed in the name of development demand the large-scale demolition or outright obliteration of the built environment. My ethnographic research in and around Yusufeli found that the town's residents tried to become not only wage-laborers as construction workers, drivers, or security personnel in one of the innumerable dam, road, and tunnel projects that mushroomed in the Çoruh Valley especially after 2010. Many of them also sought to tap into their surroundings' devastation in order to bring to life their own visions of accumulation and commodification. For some, this meant investing in or creating property with the purpose of making a profit out of the not-as-yet fully fixed compensation economy. Others opted for, and are still busy with, turning the artifacts and representations of the vanishing present, as well as the emotions they elicit, into commodity forms for future consumption. To these residents, the predicament to survive the worst effects of dispossession and displacement became less and less distinguishable from the desire to speculate on the valley's submergence.

The construction of the three large projects located in the middle part of the Çoruh Valley—the Deriner, Artvin, and Yusufeli Dams—

commenced several years apart. After long delays, the first two dam projects were completed in 2012 and 2016, respectively, and the latter could officially begin in early 2013. The more or less consecutive damming in this section of the river, followed by the start of countless road, tunnel, and viaduct projects, offered the possibility to find construction jobs whereas the payment of compensation to those villagers who lost their properties to dam lakes and building sites created new opportunities for investment. At the same time, local residents' participation in the creative destruction of land became conditional on their active engagement with how the demise of built environment and its surroundings spatially and temporally unfolded. Residents often needed to estimate the time span from the completion of one project to the start of another so that they could time their investments, for example, getting in or out of the booming real estate market in the Yusufeli town center. This made the temporal and spatial structures of destruction become co-extensive with those visions and projects of investment to elicit powerful meanings, uses, and evocations in the valley. My intention in this book, then, is to build on the recent anthropological interest in ruination (Navaro-Yashin 2012; Stoler 2013; Gordillo 2014; Tsing 2015), to ask under which conditions it is articulated with the strategies of accumulation and commodification, and to outline what sort of politics these two movements produce and help to make visible on the local level.

Second, and related to this point, hegemony is made and unmade in these efforts to prepare for and chart the uncertain future in monetary terms. I argue that the success of conservative-nationalist political projects based on economies of construction and destruction relies partly on the management of ordinary residents' attempts to endure and make a profit out of capitalist development's effects of destruction. Although the strategies and visions for remaining viable after the submergence emerged for the most part from below in Yusufeli, they operated and flourished strictly under the patronage of a political party; that is, the AKP, which has increasingly come to represent the state on the local level in the past ten years. As we will see, unlike other parts of Turkey where force is routinely deployed to dispossess and displace especially ethnically and politically marginalized groups, here municipal party networks, in coordination with prominent figures from the party, become instrumental in securing and advancing the investments of those residents loyal to the party and its politics. Such interventions constitute a relation of consent different from the one often discussed in the context of redistributive welfare measures such as social assistance programs or cash transfers that aim to garner political support among the poor.[6] They rather suppose and embody the figure of an entrepreneurial subject consolidated at the juncture where the intimate bonds of community, colocality, and nation meet the productive use of legal ambiguities and exceptions for creating economic incentives.

Thus, this book argues that strategies of accumulation and commodification, made possible by the temporalities and spatialities of infrastructural development and orchestrated through the interventions of the party-state, are what is behind the interruption of politics in a resource frontier. Interruption of politics means something different than the exhaustion or suspension of the political. It describes the constant movement by which the embryonic form of critique transforms into political organization and articulation to once again crumble back into contradictory ways of feeling and thinking. Tracing the patterns of motion between critique and consent in its different guises in the following chapters will help us to better understand what relations and structures are put in place in order to govern people without resorting to violence.

In each chapter of this book, I lay out another piece crucial to completing the puzzle of why Yusufeli's residents shifted from opposing capitalist development's effects of dispossession and displacement to, what Rosalind Morris (2008) describes in another context as, "accommodating to ruination through investment." This shift, along with the tensions and contradictions it reveals, took place under the specific circumstances by which the "capital-nation-state" assemblage (Karatani 2014)[7] behind extractive projects and logics has undergone important transformations in Turkey, as elsewhere, in the past two decades. It is this context to which I now turn to conceptually expand on the aims and arguments introduced in this section.

Bulldozer Capitalism in the National-Local Nexus

I propose a term to describe the system of power, profit, and hegemony that comes to be formed and reproduced through the destruction and recomposition of the physical environment. I call it "bulldozer capitalism."[8] I invoke the figure of a bulldozer as a metaphor slightly unconventionally here to draw attention not only to what vanishes but also to the forces and processes that connect it with what is built and created. After all, the function of a bulldozer is not limited to demolishing things. This vehicle, as I observed many times during my fieldwork research, is also used to prepare the ground for construction, opening roads blocked by landslides, or transferring sand and building equipment from one place to another. By stressing on the relationality between the bulldozer's twin uses, my intention is to recognize destruction and construction, and annihilation and accumulation as mutually constitutive processes. This dialectic perspective, I argue, can help us see both the negative and affirmative forces that constantly shift and transmute the borders and content of politics under contemporary capitalism.

Bulldozer capitalism is of course not new in Turkey. The first substantial highway networks were built and the tourism industry developed during the Cold War after the country was picked as one of the laboratories for the testing and implementation of the US-style modernization theories (Adalet 2018). Around the same time, the first large dams such as the Sarıyar and Seyhan projects were completed by the General Directorate of the State Hydraulic Works (DSI), with the technical support and financial aid they received from the World Bank increasing in the following years (Dissard 2017). With the liberalization of the Turkish economy in the aftermath of the military coup in 1980, construction projects gained pace especially in big cities such as Istanbul which were then getting closely integrated into world markets (Keyder and Öncü 1994). The export-oriented growth strategy adopted to produce cheap consumer goods for the global North, accompanied by short-term capital flows that rose steeply thanks to the early financialization of the economy, transformed cityscapes in drastic ways (Keyder 2005, 2010). Loans made available through international agencies to large municipalities in the 1990s were used to provide an infrastructural basis for the expanding managerial and entrepreneurial classes, and their nascent consumption patterns. Older and poorer neighborhoods were either demolished to open space for new roads or fell under the sway of waves of gentrification as shopping malls, gated communities, and financial districts began to dot the map to facilitate and reflect deepening class inequalities.

AKP came to power for the first time shortly after an economic meltdown, prompted by a banking crisis in 2001, brought an end to Turkey's first wave of neoliberalization. Emerging out of a crisis within political Islam itself, which was resolved by the adoption of the new party's protagonists of a market-oriented and pro-NATO outlook in place of their earlier partially anti-systemic views (Tuğal 2009), it quickly filled the political vacuum formed after center-right and center-left parties were electorally punished for their austerity politics. AKP no doubt went further than any other political party in the country's history in implementing a program of economic liberalization that included the privatization of state enterprises, flexibilization of labor markets, lifting of agricultural subsidies, and the implementation of a tight monetary policy. Yet, its effects of indebtedness, precarious work conditions, stagnant real wages, and the weakening of working-class organizations could be softened and masked through a dual strategy (Akçay 2018). On the one hand, the party's championing of democratic reforms in its early years against the Kemalist establishment within the army and the judiciary, crystallized in the opening of official negotiations for Turkey's full membership in the European Union in 2005, found support among liberal and left-liberal middle classes, and ethnic and religious minorities. On the other hand, its expansion of a series of populist welfare measures including state-provided health, insurance, and retirement benefits consolidated its power on poor and lower-middle class groups. Reinvigorating the Turkish right's deeply seated neo-Ottomanist colonial ambitions in the region in the aftermath of the Arab Revolutions and utilizing its Sunni-Turkish ideological kernel to build new alliances with ultra-nationalists and Kemalists in the domestic political scene, it gradually built after 2011 a party-state that forcefully punishes any kind of dissent in the country.

It was the global economic conjuncture of the 2000s, however, that arguably more than anything else became conducive to the AKP's political and electoral success in making possible a regime of accumulation built almost exclusively around energy, infrastructure, and real estate projects. Already after the 2001 dotcom crisis, the US Federal Reserve's decision to cut interest rates had resulted in boosting the inflow of foreign capital to "emerging markets" such as Turkey where the return of investment was much higher than in the North. Its injection of large quantities of money into global markets through three rounds of "quantitative easing" following the 2008 subprime crisis further expanded the availability of global liquidity. Even if these capital flows became increasingly short-term especially after 2008, they nonetheless allowed national governments from the global South to embark on large-scale construction projects thanks to the appreciation of national currencies, the upsurge of foreign reserves, and the decline of borrowing costs. Based on these observations, Yahya Madra and Ceren Özselçuk (2019) make the compelling argument that the financial internationalization of the 2000s provided a fertile ground for nation-states such as Turkey to entrench their sovereignties by actively intervening in and managing the economic domain through major infrastructure projects. They give the name "sovereign corporation" to those meta-market actors, such as Turkey under

Erdoğan, India under Modi, or the Philippines under Duterte, who unleash the legal and extra-legal capacities and powers of the nation-state to facilitate the extraction of surplus value, especially from land and natural resources.[9] The use of sovereign exceptions by the government, the ruling party or its leader for capital accumulation as such entails running the state like a corporation, directly deciding on which factions of capital will have the privilege to appropriate nature, resources, or labor power. The global financial expansion of the 2000s, in that sense, was not simply about finding a spatial fix for the overaccumulation crisis witnessed in the global North since the 1970s (Harvey 1982, 2010). It also paradoxically created the conditions for the spread around the globe of a political form often referred to as authoritarian capitalism, neoliberal populism, or illiberal democracy but which perhaps can also be understood as "neo-mercantilism" (Moore 2015; Madra 2017).

This is the context—the systemic crises of the US economy, the availability of liquidity that the efforts to save the US economy from falling into depression afforded to the global South, and a powerful capital-nation-state assemblage that these interventions helped to consolidate—that explains why bulldozer capitalism in its latest incarnation became the primary model of economic growth and the background against which new hierarchies and injustices, along with new needs and interests, proliferated in Turkey, as elsewhere, in the early twenty-first century. What this account overlooks however, at least in the context of large dams, is that the shift to an accumulation model under which "capital's sovereign exceptions" (Madra and Özselçuk 2019) has increasingly become commonplace is, at least in part, an outcome of transnational struggles of the previous decades. The active opposition by local communities in the global South and their allies in the North to the construction of mega-dam projects built and funded by international consortiums and organizations in different parts of the world compelled the withdrawal of international capital and the partial transition to a national regime of dam planning, finance, and construction. In Chapter 1, I recount this turbulent history of global anti-dam struggle in the 1990s and its partial success in holding international donors such as the World Bank accountable to show its implications for the rise and decline of the campaign against the Yusufeli Dam project, and the new conditions under which this project has been revived.

Anthropologists writing on capitalism and capitalist development in the past two decades have prioritized the study of those settings where transnational organizations, corporations, and NGOs come to supplant the capacities of the nation-state, inspiring phrases such as "transnational governmentality" (Ferguson and Gupta 2002) or "graduated sovereignty" (Ong 2000). Similarly, the "global-local nexus" has been proposed as the essential scalar constant through which one can make sense of the uneven, messy, and patchy character of contemporary processes of accumulation, and the social and cultural antagonisms that they make visible. Indeed, coming from different methodological and theoretical backgrounds, Anna Tsing (2005) and Don Kalb (2017), for example, seem to agree that "frictions" that arise out of international capital's interactions with the situated experiences, desires, or fears of people provide a fertile ground for ethnographically exploring the arrangements of power, class, and identity. I do not disagree with them. But I suggest that other scalar constructs deserve our anthropological attention as well. The "national" and the effects that it produces at the local level constitute a crucial scalar construct that enables accounting for the everyday social and political changes that take shape under the economies of construction and destruction that I investigate in this book.

This does not mean to imply that global flows of capital, knowledge, and politics have ceased to influence the forms of capitalism that we live with in Turkey. Nor do I suggest that we are back in the time of import-substitution driven national developmentalism. Rather, I propose to focus on a contemporary historical tendency by which "the national" in its economic, social, and ideological inflections reemerges as a powerful outcome of global fluctuations and their forces and tensions. It is more than a mere coincidence that a widely repeated slogan that the AKP chose to accompany a broader ideological offensive launched in the early 2010s precisely captures this tendency: "national and native/local" (*milli ve yerli*).[10]

Seen through the prism of this "national-local" nexus, bulldozer capitalism should be recognized as a combined and uneven model of development. Massive real estate and infrastructural projects planned and implemented both in urban and rural settings since the AKP's meteoric rise to power in 2002 are contingent in particular on spatial and regional variations marked by differences of class and ethnicity, and historical legacies of violence. For example, in big cities such as Istanbul, several lower-class neighborhoods were systematically reduced to rubble and then rebuilt as luxurious sites for housing and consumption as part of urban regeneration schemes undertaken by the Mass Housing and Public Administration (TOKI). While the predominantly Kurdish, Alevi, or Roma residents of these neighborhoods were resettled in TOKI-built houses in urban peripheries to be turned into debt-encumbered homeowners through long-term mortgage arrangements, wealthy individuals, companies, and foundations, several of which have close ties to the AKP and the Erdoğan family, made fortunes. Initially established as an administrative unit with a mission to provide housing for poor families, TOKI thus became the main instrument under the AKP regime by which profits were squeezed out of former squats and shanties, and suspicious or unruly communities are governed through indebtedness (Kuyucu and Ünsal 2010; Glastonbury and Kadıoğlu 2016). Around the same time, many middle-class homeowners of Istanbul participated in this real estate bubble by buying second or third homes in one of the recently built gated communities or handing their properties to private contractors with the intention of erecting fancier apartment buildings, even if they continued to be politically critical of the AKP governments (Bayurgil 2019). In several Kurdish cities and towns, by contrast, it was a counterinsurgency campaign launched by the Turkish state in 2015, after it terminated its negotiations with the Kurdish political movement,[11] which brought about a severe ruination in terms of its human toll and the scale of material devastation. In addition to about two hundred lives, including those of civilians, which perished during the clashes, entire settlements such as Diyarbakır's historical Sur district were nearly erased from the map within weeks. This unprecedented wave of destruction was quickly followed by an announcement that a TOKI-led construction and regeneration project would immediately commence in these areas. After urgent expropriation orders were issued for the remaining buildings, their surviving residents were pressured to either sell their houses below market prices or to purchase a new home with mortgage payments in one of the new mass housing complexes (Glastonbury and Kadıoğlu 2016; Şen 2017).

Governing along the Çoruh Valley

In the eastern Black Sea region where I conducted my research, bulldozer capitalism essentially unfolded through what can be best described as a cement-megawatt complex historically rooted in the state elites' unfailing but never entirely satisfied desire to exploit "unused" water resources.[12] Readily subscribing to a vision of "full colonization," in the words of one early observer of Turkey's hydropower potential (Nestmann 1960), while addressing the more recent problem of reducing the country's growing current account deficit exacerbated by Turkey's dependence on energy imports such as oil and natural gas, the AKP embarked on establishing a large network of hydroenergy infrastructure aimed at harnessing the entire water system of the region for electricity production. The rush to energy in the form of small-scale HEPP constructions, which gained pace by the end of the 2000s, became central to the AKP's hegemonic strategy of tailoring an image as "an able service provider, chaser of energy independence and a business-friendly sovereign" (Erensü 2018: 33). It also served to render a faction of capital subservient to the authority of the party and its leader by providing lucrative investment opportunities through a series of legal changes enacted to further liberalize the production, provisioning, and distribution of energy. When faced with resistance from local communities, the party sought to remove the barriers in front of this accumulation model by resorting to a rarely used legal procedure called "urgent expropriation" (*acele kamulaştırma*). Initially agreed upon by the cabinet of ministers and then solely left to the hands of the president after the constitutional reforms of 2017, the decisions concerning the expropriation of land planned to be used for the construction of the HEPP projects gradually became the most important strategic tool in cases of energy-related investments (Alp Kaya 2016; Erensü 2018).[13]

Figure 0.1. View of the Yusufeli town center, 2013. © Erdem Evren.

My ethnographic material comes from the middle part of the Çoruh Valley, especially the town of Yusufeli and its villages, where I conducted fieldwork research for a total of twelve months between 2012 and 2018. It became clear during my initial visit in 2011 that the heavy damming of the Çoruh River since the early 1990s had given rise to a context markedly different from other parts of the region. The remarkable scale of material devastation and displacement authorized by its planners and the vicissitudes of planning and finance to which it became subject over decades had turned especially the Yusufeli Dam project at first sight into a peculiar example of infrastructural development. But while Yusufeli is distinct in many ways, it is however not an aberration. It should perhaps be seen as an "abnormally normal" (Ferme 1998) case that illustrates how sovereign exceptions can work both in constraining and productive ways to further the interests of capital and the state within the national-local nexus around which bulldozer capitalism has been reconstituted.

In Chapter 3, I discuss in detail how the planning of small-scale HEPP and mining projects and the implementation of cadastral surveys enabled the expropriation of private land and the enclosure of commonly used meadows and forests in this part of the valley as well. I explain that in some of the villages in Yusufeli, urgent expropriation orders were issued by the cabinet to bypass the popular opposition against the construction of a series of small hydropower projects. Similarly, I emphasize the role that legislative changes and presidential decrees played in keeping construction companies, including the domestic consortium behind the Yusufeli Dam project after 2010, financially buoyant by constantly feeding them with new tenders or periodically canceling their debts. At the same time, I observe a different modality of governing which can be considered productive insofar as it serves to enlist the local community to take part and speculate on the destruction of its built environment and ways of life. I found that the party-state, through its elected officials such as the mayor of Yusufeli and its local branch, deployed temporary and contextual legal arrangements, which Partha Chatterjee (2004, 2008) finds vital for the governing of the subaltern in the contemporary moment, to secure or advance the investments of especially those residents loyal to the party and its politics.

Chatterjee's work is a contribution to an important debate among scholars writing on the fate of communities marginalized as a result of mega dams, special economic zones, and other extractive and infrastructural projects in India and it is therefore worth revisiting here in order to make sense of what is particularly novel about the nature of the AKP's dealings with Yusufeli's residents. Building on Kalyan Sanyal's (2007) seminal work, Chatterjee (2008) argues that the normative commitment of the Indian state to rural communities necessitates the reversing of the effects of land grabs through the partial rehabilitation of its victims. This commitment stems from the reorganization of modern statecraft around the principle of legitimization attributed by him to the Indian nation-state's adoption of a democratic system based on universal suffrage. Because national political parties rely on the votes of the communities that they seek to displace, the use of force remains a limited option for governing rural populations. Instead of violence, he puts the emphasis on those procedures of governmentality that provide welfare and security to subaltern classes. Appealing to ties of moral solidarity, subaltern communities actively mobilize behind collective claims to influence the implementation of administrative policies especially concerning property relations. Many of these policies bend and stretch existing rules and regulations to make room for arrangements that benefit these communities. In place of popular sovereignty and its promises of equal citizenship epitomized by developmental interventions, Chatterjee (2004) therefore sees the mobilization of the dispossessed in the political society for some form of compensation or welfare benefit as the main mechanism through which the subaltern takes part in its governance at a time when they are increasingly stripped of their land and other means of subsistence.

Yusufeli's residents can be described neither as peasants nor subalterns. They do not easily fit into the categories of civil society comprised of an educated middle class or the political society formed by displaced rural residents and marginalized urban settlers either. More importantly, even if the locals' electoral power and conservative-nationalist convictions may have emboldened the party-state's normative commitments to address their grievances, as Chatterjee would argue, its use of paralegal arrangements around property does not by any means serve to reverse the effects of displacement and dispossession. As his critiques (Baviskar and Sundar 2008; Levien 2018) have also argued, these interventions did in effect more to enrich a handful of local politicians and property owners than to find actual solutions to the community's social and economic misfortunes.

In Chapters 2 and 4, I argue that it is the party-state's inculcation of expectations among residents to make deals over and capitalize on the land, property, and social relations made bound to disappear by the coming submergence that becomes crucial for building a relation of consent. Formal and informal forms of bargaining allowed by local authorities opened the possibility of addressing, at least in rhetoric, the plethora of material and immaterial grievances facing the local community. Paralegal arrangements carefully worked out by the mayor and his office, on the other hand, mobilized them to tap into the destruction of their surroundings and livelihoods by creating economic incentives tied to the construction economy and the compensation schemas. I consider this powerful incitement to negotiate and invest in ruination an understudied yet crucial aspect of how authoritarian regimes continue to garner support among those communities on whose livelihoods they wreak social, economic, and environmental havoc. One polemical point that this book therefore intends to make against some of the recent ethnographies on Turkey is that it is not simply religion or nationalism but the material relations, interests, and needs woven around infrastructure, real estate, and resource extraction by conservative-nationalist regimes that underpin the making of hegemony. This, I argue, would not have been possible under the international regime of dam finance and construction under which multinational construction and energy companies operate.

Nor would it be conceivable in another place where such close ties between the local community and the national government are absent. One powerful myth that my interlocutors in Yusufeli repeated to me is that Yusufeli is the birthplace of the AKP. After completing his prison sentence for having read, in one of his political campaigns in 1997, a poem deemed incendiary by Kemalist judges, Recep Tayyip Erdoğan came to stay, according to this narrative, in the family house of Kadir Topbaş, the future mayor of Istanbul, in the neighboring Barhal Valley to start the preparations for launching his new party. Photos taken with Erdoğan himself around this time were still adorning more than a few shops that I frequented during my research. Similarly, the presence of cabinet ministers and high-ranking bureaucrats with ties to Yusufeli was regularly mentioned as a source of pride, as well as a marker of proximity to central power and authority, by the town's residents. Their affective investments in the real and imagined connections between the local and the national, mediated by politicians and administrators recognized as co-locals, were crucial in cultivating hopes for negotiating the terms and conditions of their displacement and dispossession. As one interlocutor, a young member of the cultural association, once told me: "Because of Kadir Bey and others, people thought their social and economic grievances would automatically be addressed, that they could talk through their issues, problems and expectations with them."

Governing in the interests of capital accumulation by appealing to sensibilities of community, nation, and co-locality attests to the significance of those material and immaterial bonds, imaginaries, and activities that sustain these relations. Taking my cue from political theorists who warn against the dangers of recognizing commons invariably as the locus of resistance against the state and capital—for example, David Harvey (2011) who reminds us that gated communities too are an example of commons or even Michael Hardt and Antonio Negri (2009) who speak of the family, the corporation, and the nation as corrupt forms of the common—in this book I use the term "conservative commons." My intention is to draw attention to the role that an assumed common identity, based on a shared past and place of origin, along with similar religious beliefs, ideological inclinations, and ways of life, plays in particular in fostering trust in the present and in building anticipation for the future. In some ways, conservative commons come close to describing those ties of moral solidarity to which subaltern communities, in Chatterjee's (2008) formulation, appeal to make certain demands from the state. However, in the majority of situations described in this ethnography, such ties turn into the material of unfulfilled promises made by the party-state and its local representatives.

Time, Space, and Ruination

One key conceptual category that holds together the vexed relations articulated in this ethnography between ruination, accumulation, and dispossession is temporality. The spatial effects of infrastructural projects are partly contingent on the durations, tempos, and iterations with which the processes of construction and destruction unfold in any given location. As previous ethnographic works have shown, these temporal frames are crucial for how people materially and conceptually approach, emotionally experience, and politically respond to the changes in their surroundings and livelihoods (Abram and Weszkalyns 2011; Baxstrom 2011; Laszczkowski 2011). This book is foremost concerned with a situation in which several members of the target community endure infrastructural devastation and wait for their involuntary resettlement for almost their entire lives. This protracted ruination, I argue, is crucial for understanding the changing political responses to capitalist development first by intensifying the residents' everyday forms of discontent towards dam constructions and later constituting the ground upon which they seek to realize their own projects of accumulation and commodification.

While the building of almost all dams planned as part of the Çoruh Energy Plan had to be deferred either for financial or geostrategic reasons, it was the Yusufeli Dam project that witnessed the longest delay, not least as a result of the residents' own successful anti-dam/anti-displacement campaign for almost a decade. The repeated interruptions in the realization of this project gradually bolstered a shift in the dominant subject position in the town from resistance to resignation. Its residents increasingly assigned monetary value to the organic and inorganic life in the valley by making an effort "to estimate the likely time of its expiry" (Morris 2008: 205). This effort placed ruination in the same temporal structure with the strategies of accumulation to decisively defeat the earlier anti-displacement activism in the town and created the conditions of possibility for conceiving the past, present, and future in unusual constellations: creating property in the present to be able to profit from its demise in the future or turning the remnants of the past and the vanishing present into commodities with the intention of selling them in the future. In this unpredictably expansive resource frontier, it seems as if *time itself* becomes yet another resource that its residents attempt to appropriate from below.

In Chapters 2 and 4, I consider "looking forward," a term that came out of my conversations with my interlocutors, as the key temporal orientation that structures the social, economic, and political responses to the protracted nature of ruination. Rather than approaching it as a merely subjective sensibility concerning infrastructural transformation however, I stress that it took shape at a particular moment under the specific conditions of the Yusufeli Dam projects' renationalization and the AKP's victory in the municipal elections. "Looking forward" could begin to name the anticipations of the residents only after the party-state managed to steer the uncertainties growing out of the devastation of the valley and its residents' livelihoods. By delivering the old paternalist promises of employment through party networks, on the one hand, and by making paralegal interventions concerning resettlement and the future compensation economy, the AKP successfully turned some of bulldozer capitalism's victims into its laborers and entrepreneurs. This is then what essentially qualifies sovereign exceptions as a productive or enabling governing strategy: they target and manage people's sensibilities of time and temporality as opposed to securing their immediate exclusion from space.

Recent ethnographies (Harms 2013; Bear 2015) that follow the lives of ordinary people caught up in turmoil caused by large-scale infrastructural changes emphasize the emergence of entrepreneurial subjects. This figure is central for my work as well. Owing to the specific conditions rooted in the protracted nature of ruination and the power of intimate relations of community and co-locality, the strategies of Yusufeli's residents for surviving the worst effects of destruction and displacement gain a business-like quality as they make decisions about their lives based on cost-benefit analysis. This is as much true for ordinary workers who need to calculate how long they can continue to work at a particular building site and when they can begin a new job in another one, as for those residents who make or invest in property to make a profit out of the future compensation economy. But this entrepreneurial urge carries strong elements of speculation generated about the disappearance of land, community, and the past, which philosopher Michel Feher (2018) recognizes as a characteristic of subjects fashioned by financialized capitalism that he intentionally, and perhaps a bit too neatly, distinguishes from neoliberal subjectivity.

There is, however, another sense in which the residents are compelled to make calculations about the effects of destruction and expropriation across different temporalities and spaces in this resource frontier. In Chapter 3, I explain that as they continued to wait for the construction of the Yusufeli Dam project, Yusufeli's villagers also experienced the more or less simultaneous implementation of mining and HEPP projects, and cadastral surveys. Their effects of dispossession became spatially entangled with one another to give rise to different political responses: resistance against small hydropower projects, resignation towards the transfer of the commons and private property to the Treasury and the General Directorate of Forestry, and recognition of gold and copper mining as the only viable economic opportunity after the submergence.

The contrast between the desires and fears that different extractive technologies fuel in and around Yusufeli highlights that the hegemonic project by the party-state is never entirely complete. The embryonic critique embedded in the common sense continues to electrify new moments of struggle and protest, even if they once again become blocked or get interrupted. I therefore fix my ethnographic gaze also on the cracks, as much as on the connections, between the national and local scales to be able to understand some of the new social and political forces and conditions that still make opposing the party-state and private capital's projects of extraction possible.

Violence, Its Reverberations, and the Researcher

On the second day of Bayram (*Eid-ul Fitr*) in August 2012, Orhan, my occasional driver in the field, brought me to the villages half-flooded by the reservoir of the Deriner Dam, by then the most gargantuan project completed on the main tributary of the Çoruh River. On our way, he pointed out to me, one by one, the locations of a gas station, a school, a cemetery, and several olive groves, as well as the itinerary of the 1994 World Rafting Championship—all of which were now hardly discernable in the opacity of the river-cum-lake. The view from the hilltop where we parked our car was immediately breathtaking and disquieting at the same time. Two dozen houses scattered around the shades of green were disfigured, as ant-like shadows below us were industriously cutting, pilling, and loading stuff into trucks. The buzzing sound of chainsaws moving around the ruins made Orhan break into a laugh, an ironic one expressing not only sadness but also anger. "The villagers are removing the wooden parts of their houses before their village gets entirely flooded," he explained. "It's often juniper, very valuable." He then slowly turned his eyes to the shores of the village, where the land was being swallowed bit by bit by cloudy wa-

Figure 0.2. The village of Sirya, Artvin, half-flooded by the Deriner Dam reservoir, 2013. © Erdem Evren.

ters and where the white minaret of the village mosque had already visibly shrunk. "Our grandfathers and great-grandfathers built this place," he continued. "Now these men are taking everything with them, as if we're at war. And they're doing it on a religious day like this. This isn't something to be done today."

Twenty minutes later, we were driving back to Yusufeli, where Orhan was then working as a temporary rafting instructor and I was asking its inhabitants how they see their own approaching displacement or "coming annihilation" (*yaklaşmakta olan yokoluşumuz*), in the words of one interlocutor. Having retreated into my dreamy thoughts as we continued to traverse the valley, I recalled the scenic village in which Orhan had earlier hosted me in his small garden by the river. I asked him if he, too, just like the villagers that we just saw, was planning to sell the ruins of his house before its disappearance. "Yes," he replied, abruptly and visibly uncomfortable. I thought I saw his body slightly twitch. "Yes. We'll do the same thing."

Orhan's words and embodied reactions, or more precisely my ethnographic reconstruction of them here, reveal an evident tension between attachments to and commodification of the built environment amid its obliteration. I witnessed this tension under various guises in numerous other occasions in Yusufeli: its residents often spoke of the ceaseless material destruction in the valley as some kind of injury and explicitly named the scarring of their relation to the landscape and its past as suffering even though they try to make a profit out of this destruction or continue to attribute monetary values to its ruins. The demolition of a landmark such as a bridge or the submergence of a village in another part of the Çoruh Valley because of dam or road constructions caused genuine suffering for Yusufeli's residents, foretelling the coming disappearance of their town and the dissolution of their community and ways of living. This tension continues to afflict social relations with affects such as disturbance and grief. In this book, I explore it as a generative force that embroils infrastructural violence in other social and political injuries—some immediate, others forgotten or disremembered.

This brings me to the final, and in some ways the least expected finding, of this ethnography: while bulldozer capitalism in this part of Turkey operates predominantly as a relation of consent predicated on the incitement to negotiate and speculate on ruination, violence remains its invisible background condition and a force around which some of its contradictions become crystallized. I found that at least some of the property and agricultural land that the residents own and invest in to receive higher compensation payments once belonged to the Armenian residents of the valley who became subject to mass deportation and killings in 1915. In Chapter 5, I trace this partly forgotten and to this day vehemently denied regional episode of the Armenian Genocide in my conversations with the current residents, and in my review of historical records, survivor memoirs, and a small but representative cache of archival material that I serendipitously stumbled upon during my research. By bringing out this episode of destruction and dispossession, my aim is to highlight how large-scale expropriation of wealth in the past continues to reverberate in the present. The figure of the dead Armenian that the residents curiously invoke to make sense of the coming displacement of their community and the devastation of their built environment, I also show, permeates the relations between ruination and accumulation with unusual meanings and affects.

This book therefore proposes to go beyond the here-and-now of large-scale extraction of resources to shed light on in what ways this process is entangled with and embedded within previous acts of violent dispossession. Marx ([1887] 1977), as it is well known, devoted the last section of *Capital Volume I* to the study of "the so-called primitive accumulation" that provided the initial capital and labor power necessary for inaugurating the shift to the capitalist mode of production. While his vivid depictions of enclosures, mass murder, plunder of raw materials, and penalization and incarceration in England were meant to explain the originary process by which direct producers were forcefully divorced from their immediate means of production to fill in the factories as "free laborers," these passages have received renewed attention in recent times to rethink the role extra-economic forces play in relation to and alongside capital accumulation under contemporary capitalism. David Harvey (2003), for example, coined the term "accumulation by dispossession" to mark how force is deployed to resolve the ongoing crisis of overaccumulation under the guises of financialization, urban renewal projects, and indebtedness. Other authors writing more directly on capital's extractivist logics and operations (Fraser 2014; Moore 2015; Mezzadra and Neilson 2019) draw our attention to the disparate ways in which violence remains the connecting tissue between expropriation and exploitation.

In trying to understand why the target community of a mass political violence that took place more than a hundred years ago is remembered in the context of contemporary processes of resource extraction, displacement, and dispossession in the Çoruh Valley, my work follows Harry Harootunian's (2019) recent intervention that urges us to consider primitive or originary accumulation not as a one-time event but a process that continues to unfold across different spaces and temporalities down to our present. Mainly concerned with the tragic legacy of his own parents who as survivors established migrant livelihoods in the US while suppressing every trace of their previous lives in Anatolia and the memories of the genocide, Harootunian proposes to look at how the destruction of Armenian everyday life continues to reproduce the society in its different material and affective registers. Yet, this point seems to me as relevant for the Turkish-Sunni community of Yusufeli as for the Armenian diaspora in the US and elsewhere. Building on the various narratives of my interlocutors on the displacement of the valley's former Armenian residents and the expropriation of their property at a time when they face their own imminent resettlement and dispossession, I discuss the ways in which the speculative forms of accumulation in contemporary Yusufeli are materially and discursively shaped and complicated by the originary violence that occasioned the originary accumulation in the valley. In that sense, I intend to draw attention to another layer of the dialectic relation between destruction and construction, and annihilation and accumulation, which I argue gives bulldozer capitalism its defining character.

It would perhaps be helpful to remember here the original meaning of the word "bulldozer": "a person who intimidates with violence" (Bellér-Hahn 2014: 188).[14] While I did not feel intimidated almost at any point during my research, my ethnography among the conservative and nationalist Sunni-Turkish residents of Yusufeli, and gradual understanding of their complicity, involvement, and complaisance of different instances and episodes of political violence requires me to reflect on my own positionality.

What brought me back to Yusufeli to conduct long-term ethnographic research was the desire to understand why its anti-dam/anti-displacement campaign gradually went into decline after successfully preventing the construction of the dam project for more than a decade. Perhaps more decisively though, I arrived with a sense of curiosity about what it means to lead a life in this piece of land squeezed between steep rocks and the river's tributaries without much arable land, yet still exuding abundance and charm thanks to its Mediterranean micro-climate, which had long ago been made destined to disappear by the state. The flat that I moved to with my partner and son in 2013 was a sublet from Ahsen;[15] one of the two women based in Germany who got involved first in the Yusufeli campaign and later in the efforts to mobilize the villagers against the plans for the construction of HEPP projects. In our first meeting at a café in Berlin, she carefully listened to my proposal to conduct long-term ethnographic research in Yusufeli and promised to inform her contacts in the town about my arrival. Her friend, Zeliha, was a key activist who coordinated the international part of the campaign for several years, closely researching the international construction consortiums showing an interest in the Yusufeli Dam project, and using her contacts from various European NGOs to put pressure on export credit agencies (ECAs) in order to prevent the release of funds. Zeliha remained in close contact with three key anti-dam activists from the town who, after some initial hesitation, allowed me to stay in the town as an ally of the struggles against hydropower projects.

The former head and lawyer of the local association, Ragıp Bey, was, at the start of my research, a member of and a local candidate from the ultra-nationalist Nationalist Action Party (MHP). Muhsin Abi, the founder of the town's only photo studio, was a militant of the MHP's paramilitary wing, the Grey Wolves movement, in his youth and subsequently worked closely with the Grand Unity Party (BBP), which is an Islamist-leaning offshoot from the MHP. The leaders and militants of the Grey Wolves, and these two political parties in which they took part, have been directly responsible for killing hundreds of socialist activists and organizing pogroms against the Alevi minority in the 1970s.[16] To this day, they remain as the most outspoken ultra-nationalist political forces in Turkey. After both MHP and BBP decided to support Recep Tayyip Erdoğan in all the crucial turning points in the most recent history of Turkey, including the referendum on the transition to a presidential system in 2017, and eventually formed electoral alliances with the AKP, Ragıp Bey and Muhsin Abi left to found the Yusufeli branch of the Good Party (İP), a new nationalist party catering to secular Turkish-Sunni vote disillusioned by these two parties' alliance with the AKP. Hikmet Abi, on the other hand, is a retired policeman who used to head an anti-terror unit in the neighboring city of Erzurum in the 1990s when systematic torture and extra-judicial killings of Kurdish and leftist activists reached its paramount. Visibly marginalized politically for having remained staunch opponents of the AKP at a time when the party was successfully swallowing a wide range of right-wing constituencies in Yusufeli as in other parts of the country, these activists continued to resort to a nationalist-conservative rhetoric in order to animate the earlier mobilization against the dam project in the town.

Within the local association and the core anti-dam group including Zeliha and Ahsen whose leftist politics posed a stark contrast to these activists' ultra-nationalist worldview, there was a general agreement from early on to gloss over party politics and ideological differences to join forces for saving the town from being submerged by the Yusufeli Dam project and later to organize the residents against small hydropower projects. While this worked well for a long time, it nevertheless placed me in a tricky position. Because of my close relations with Muhsin Abi, whose photo studio became one of my primary sites of research where I spent countless hours catching up on the latest developments about the project and got to know several people from the town, I was instinctively perceived by some of the prominent figures in the town as part of the diminished anti-dam campaign. Some of the self-declared leftists and social democrats, on the other hand, initially kept their distance from me as they thought me as an odd-looking member of this ultra-nationalist circle. Leftist colleagues and friends from Turkey too expressed their discomfort about the presence of ultra-nationalist figures as anti-dam activists in my ethnographic account when they heard me presenting parts of my work over the years. A few of them even outright criticized me for "doing research with fascists." Having grown up in a leftist family, I also found some of these intimate exchanges with my closest interlocutors emotionally challenging and confusing. I deeply cherished the father-son relation formed over the years between Muhsin Abi and myself as we smoked one cigarette after another in front of his photo studio, something that the conservative morals of the small town prevented him from doing with his own son, while chatting about the latest gossip in the town and politics in general. Other times, I felt a visceral revulsion by the sight of him or Hikmet Abi taking my son into their arms.

Doing research in Yusufeli in close contact with these figures and many others proved to be a lonesome and challenging yet instructive experience. Whereas my own positionality as a male researcher prevented me, for the most part, to have meaningful interactions with women living in and around the town, the growing suspicions of some residents about the aims of my research at times blocked the possibility to continue conversations or build deeper relations. In the end, I found, just like some of my interlocutors did, a history of wreckage piled layer upon layer. Walter Benjamin (1968), as it is well-known, was a unique voice among the Marxists of his generation for critiquing the idea that accumulation is the motor engine of history conventionally understood to be progressing in a linear fashion. Against this tendency of "historicism," he instead developed a sophisticated intellectual project that considers history to be based on the accumulation of the violence of the past, and the disruptive entanglements of its ruins with the present. While *Bulldozer Capitalism* does not strive to become a theory book, by any means, thinking Benjamin's insights alongside Gramsci's formulations on the politics potentially takes us beyond the immediate settings of Yusufeli, the eastern Black Sea region, or Turkey. Viewing the contemporary moment from the vantage point of a dialectic relation between annihilation and accumulation may allow us to see better in what ways the relations between consent and violence facilitate and reproduce its conjunctures of economic, political, and ecological crisis.

I read and reread the works of these two revolutionary martyrs of fascism at a time when the uprisings and mass protests in Turkey were followed by countless bombing attacks, waves of prosecution and mass incarceration, a coup attempt, occupations and invasions by the Turkish army, and the rise of an oligarchic power structure becoming each day more ruthless and para-militarized. In their words I searched, I believe, for an intellectual and moral compass to navigate through these difficult times. In Brecht's "Motto," which I use as an epigraph to this introduction, I may have found a purpose, as well as a tonality, that I hope this book could at least try to do justice to.

Chapter Outlines

By situating the rise and decline of the Yusufeli campaign within the broader changes witnessed in the planning and construction of large dams in the past three decades, Chapter 1 makes the argument that the relative strength or weakness of struggles against large infrastructural projects is subordinated to the differential power of finance. Fixing its gaze on the conditions by which the work of articulation and organization against dam building unfolded in and beyond Yusufeli, this chapter describes two related processes that help us to make sense of why local campaigns from the global South failed to decisively impede construction: the detachment of dam capital from its transnational nexus as a result of the circulation of global norms and the dependent financialization of the 2000s. Locals' reflections on and recollections of their struggle and its aftermath recounted in this chapter help us to understand some of the circumstances under which the construction and financing of large dams began to be reconfigured around a capital-nation-state assemblage. They also reveal some of the new capacities that the end of the transnational regime of dam-building bestowed on the party-state for integrating a community facing dispossession and displacement into its own vision of capitalist development.

Chapter 2 focuses on how the experiences of time and space at the local level becomes crucial for the implementation of this particular vision of capitalist development. It shows that the material expectations from and investments in a future to come, captured by the phrase "looking forward," were controlled by the AKP networks to make the residents subscribe to the ruination of their lives and built environment. By putting this key future orientation at the center of analysis, this chapter intends to highlight the class dynamics behind the rise of the party-state and the consolidation of its hegemonic project. Especially local shop owners, who become involved in the party networks and construction businesses around the same, as well as national politicians with family connections to the town, help to channel and for the most party contain ordinary people's desires and hopes regarding compensation, resettlement, and employment. These class relations illustrate the workings of a consensual political practice predicated on the capillary direction of the entire social fabric.

One of the striking features of damming in this part of Turkey is that it goes hand in hand with other extractive processes such as the implementation of prospective mining and cadastral surveying. One ethnographic issue that Chapter 3 grapples with is the residents' varying political reactions to the different aspects of this expanding material destruction and dispossession. Looking at the responses of resignation, accommodation and protest in Yusufeli and its different villages, this chapter demonstrates that political agency in environmental and resource-based disputes is formed relationally, emerging out of people's various considerations of what will be lost and what can be gained as different processes of destruction and expropriation become spatially and temporally linked. The political responses to entangled dispossessions in and around Yusufeli take shape against the background of two important developments: the convergence and divergence of a plethora of actors' interests and fears and expectations, and the transformation of rural livelihoods as a result of the dissolution of rural life and the decline in agriculture as a viable economic activity. The chapter comments on how these developments help to sustain or alternatively open cracks within the party-state's governing apparatus.

Chapter 4 continues to reflect on the cost-benefit analysis to which the residents render different aspects of their lives and built environment under the conditions of ruination, this time in the context of the economies of construction and destruction flourishing in the town since the early 2010s. It chronicles those strategies and visions for capitalizing on the land, property, and social relations which are about to disappear with the completion of the Yusufeli Dam project. From buying or creating property with the purpose of receiving higher compensation payments to developing ambitious plans for selling artifacts bound to vanish with the submergence, Yusufeli's residents invest in the loss and destruction of things and ways of living as a source of valuation. The chapter discusses the new social and economic divisions and hierarchies engendered and reproduced as a result of these speculative ways of thinking and behaving, and points to precarity and indebtedness as the common outcomes of the broader entrepreneurial drive in the town.

But what does it mean to lead a life in anticipation of its demise? To be able to answer this question, Chapter 5 turns to those narratives and practices in which the residents employ the conceptual opposite of "looking forward," that is, "looking backward" and discovers that the disappearance of Armenian lives, and the destruction and plundering of their property more than a century ago continues to reverberate in the present. The deployment of violent processes (war, annihilation) as metaphors to describe the physical damaging of the valley in its different temporal registers and the ongoing search for Armenian treasures suggest that the originary violence and its effects of dispossession continue to be constitutive of the contemporary processes of accumulation. The ghosts of the past may have reappeared to make people "look backward," the chapter ultimately speculates, because there is no longer a political agency that can prevent the loss and injury caused by capitalist development in the present.

Notes

1. Similar to dams, HEPPs, also known as small hydropower plants or *hidroelektrik santralleri* (HES) in Turkish, exploit the vertical distance between two sections of a river or a stream to convert the kinetic energy of water into energy. There are two main differences from conventional dams however: first, they lack a reservoir in which water is collected; second, their install capacity usually does not exceed 10 MW, even though this threshold has been subject to several changes by the hydraulic authorities in Turkey in the past twenty years (Erensü 2016). If the quintessential image of a modern large dam is a massive concrete wall that dissects rivers into two parts, then for a small hydropower plant it would be kilometers-long pipes inside which water is captured and then released back to the riverbed after passing through the turbines within a powerhouse.

2. In one of the earliest and long-lasting environmental struggles in Turkey, Artvin's residents organized around the Green Artvin Association (Yeşil Artvin Derneği) in 1995 to protest against the Canadian mining company Cominco and its Turkish subsidiary Inmet's plans to extract gold from Cerrattepe with the use of cyanide. After numerous demonstrations and court cases, Inmet's operating license was revoked in 2008 for environmental violations. A new bidding process that took place in 2011 however saw the deliverance of the tender to Özaltın Company, which works on behalf of the notorious construction conglomerate Cengiz Holding. After another protracted legal struggle led by Green Artvin, Özaltın's construction vehicles finally entered the Cerrattepe area in 2016 after the gendarmerie clashed for hours with thousands of people from the city and neighboring towns who had gathered to stop the project. I note this brief history of the Cerrattepe struggle here not only because Cengiz Holding later became part of the domestic consortium behind the construction of the Yusufeli Dam project, my main focus in this book, but also to draw attention to the fact that the shift from an international to a national regime of mining and its political effects in Cerattepe are somewhat analogous to the developments in dam planning and building that I contextualize and discuss in the next sections and chapters.
3. In the course of my research, I have come across conflicting figures, even in official reports and interviews with DSI (General Directorate of the State Hydraulic Works) officials, as to the exact number of villages planned to be inundated and of people expected to be displaced with the completion of the Yusufeli Dam project. This confusion is in part related to the arbitrary decision to include or omit from the list those villages that will be partly affected. In addition to the town center and three villages that will be entirely submerged by the dam reservoir, fourteen other villages will either be half-flooded or partly bulldozed to build construction sites. Half-flooding is often an equally severe outcome of damming since the loss of agricultural land or parts of the settlement poses a serious threat to the already precarious rural existence in the villages. Even though it is claimed that twenty thousand people will become subject to displacement in most newspaper reports and official statements, I found that even the planners and local politicians are not exactly certain about this figure. Despite seasonal fluctuations, Yusufeli's total population, including the town center and the villages, did not exceed ten thousand people in the second half of the 2010s.
4. Throughout this book, I use "anti-dam" and "anti-displacement" either interchangeably or together with a slash in-between. Even though I am aware of the difference between the political positions that they imply, several of my interlocutors used both terms to express their opposition against the flooding of Yusufeli and its villages for the production of electricity. At the same time, local activists, especially during their early encounters with the state authorities and national politicians, strategically framed their campaign around the issue of displacement to avoid being criticized for opposing development—an almost sacrosanct discourse and practice that has taken both the political right and left under its spell for decades (Arsel 2016a).
5. The notion of frontier, especially in its usages by politicians and state elites, often invokes "backwardness" to legitimize technical solutions (Davis and Burke 2011). Equally importantly, the taming of "wilderness," as Dale Stahl (2019), among others, convincingly argues in his work on the Keban Dam project in Southeastern Turkey, goes hand in hand with the control, assimilation, and removal of people deemed nonmodern or dangerous. Environmental engineering and social reshaping, in other words, become deeply intertwined to enact the political and technological goals of the nation-state. While such imaginings were not entirely absent in the minds of those engineers and planners who made the decision in the early 1960s to dam the Çoruh River, my reference to resource frontier here simply intends to underscore the intensity of extraction in this particular geographical location. As it will become clear in

the next chapters, a certain reassessment of the value of rural life, along with the idea that natural resources are underutilized, continues to constitute the ideological background of the projects intended on producing economic value out of nature. But what I am also interested in understanding is how these visions come to shape the desires and expectations of ordinary people under conditions connected with the rise of the AKP.

6. For the changing welfare regime and its political effects under the AKP, see Buğra and Keyder (2006) and Yörük (2012).
7. Kojin Karatani uses the term "capital-nation-state" to study the history of social formations from the perspective of modes of exchange. Each element of this assemblage, according to him, has its own distinct logic and principles but they have become deeply enmeshed in one another to the extent that "capital" or "nation-state" in itself would have little explanatory power. I borrow the term here for the less ambitious purpose of contextualizing infrastructural projects such as the Yusufeli Dam project, and making sense of their politics in the light of the interconnectedness between these entities.
8. Inspiration for the terms comes from Lovering and Türkmen (2011), who describe the state-led real estate boom and the displacement of the urban poor in Istanbul as "bulldozer neoliberalism." Adaman et al. (2014), Çavuşoğlu and Strutz (2014a, 2014b), and Eder (2015) also come close to invoking the term when they write about the speed of destruction or the coexistence of destruction with construction often in the context of urban regeneration projects in big cities.
9. Another closely related concept used to describe nation-states' increasing role in developmental projects in Latin America in recent years is "sovereign development state" (cf. Mcneish 2013).
10. The phrase *milli ve yerli* and its other variations have frequently been invoked by Turkey's leading conservative-nationalist and Islamist politicians and ideologues since the 1960s at least (Bora 2016). Its widespread recirculation under the AKP after 2015 coincided with the party's adoption of an ultra-nationalist rhetoric. In this narrative, whereas the AKP is presented as the only political party that truly defends the "national interests" and represents the authentic "people's will," others, especially the Republican People's Party (CHP) and the Peoples' Democratic Party (HDP), are marked as "foreign," "unpatriotic," or "terrorist." At the same time, Recep Tayyip Erdoğan and other AKP politicians frequently deployed this expression as a discursive marker in the opening ceremonies of several flagship projects, thus, referring to *milli ve yerli* energy, automobile, and military technology.
11. Also known as the "peace process," "solution process," or the "Kurdish opening," the negotiations that took place between 2013 and 2015 aimed to bring an end to the decades-long fight between the Turkish state and the Kurdistan Workers' Party (PKK). After the HDP's phenomenal success in June 2015 elections, which deprived the AKP of a majority in the parliament, Erdoğan made the decision to unilaterally end the negotiations, after which the clashes with the PKK guerillas and the large-scale attacks on Kurdish cities and towns resumed. Finding support from the ultra-nationalist Nationalist Action Party (MHP), the AKP blocked the efforts for forming a coalition government and once again went to the elections in October 2015, after which it regained its majority.
12. Referring to "unused" or "wasted" resources to justify their extraction and expropriation is a powerful rhetorical device adopted also by the AKP politicians especially in connection with the construction of small dam projects. Thus, in a famous speech that he gave in December 2012, then Prime Minister Erdoğan announced: "We do not have the luxury to waste our time by merely watching our river, as the idiom 'water flows, Turk watches' (*su akar, Türk bakar*) indicates. We changed this idiom, now it is 'water

flows, Turk builds' (*su akar, Türk yapar*). We must use our limited resources in the most efficient way possible and turn them into energy" (mentioned in Eren and Büke 2016).

13. Alp Yücel Kaya (2016: 79–81) calculates that a total of 1801 urgent expropriation decisions were taken by the cabinet of ministers during the Republican era, the majority of which belong to the AKP governments. While 92 percent of them concerned the investments in the electricity market, 212 of them were directly related to HEPP projects. Kaya's list covers the period until 2014 and excludes the expropriation decisions implemented on behalf of the Ministry of Finance and through Presidential decrees after 2017.
14. I would like to thank Moritz Roemer for bringing this to my attention.
15. I changed the names of all my interlocutors in this ethnography, except for the past and current mayors of Yusufeli, but I keep the form in which I address them; thus, using the informal *Abi* (brother) and *Abla* (sister), and formal *Bey* (Mr.). I decided not to anonymize place names.
16. Just to give two examples, before he created the BBP in 1993, the late Muhsin Yazıcıoğlu, whose first name I am appropriately using as a pseudonym for my interlocutor, was charged and imprisoned for taking part in the murder of several leftists, including the seven university students and members of the socialist Turkish Workers' Party in Ankara in October 1978. Another cofounder of the party, Ökkeş Şendiller (Kenger) is widely considered to be one of the planners of the pogrom against the Alevi community in the city of Maraş in December 1978, resulting in the deaths of more than one hundred people.

— Chapter 1 —

The Rise and Decline of an Anti-displacement Campaign

From Opponents to Applauders

Elderly residents living in the middle part of the Çoruh valley vividly remember to this day the first engineers who arrived from the capital in the early 1980s to conduct technical surveys on the hydroelectric energy potential of the river. A few among them even speak fondly of their time as construction workers in designated sites. This survey followed up on an energy plan hastily assembled about two decades earlier in the offices of the Department of State Planning (DPT) in the heyday of World Bank–sponsored national developmentalism. Hence, its published outcome, named the Çoruh Energy Plan,[1] announced in the best high-modernist fashion the construction of ten big dams on the main tributary of the Çoruh River with hardly any mention of their human and environmental costs.

In retrospect, this large-scale extraction plan seems to have been prepared and matured under relatively convenient conditions. The early years of military rule that coincided with the finalization of the Energy Plan effectively prevented even the mildest critique of the valley's submergence for energy production from finding popular expression. The plan itself was not made public in its entirety until the late 1990s, in order to successfully conceal the extent of the social, material, and ecological devastation its design was destined to precipitate. Until then, the earliest projects either found quiet acceptance across the political spectrum or, at best, catalyzed distinct visions for a negotiated development. Muratlı Dam on the downstream of the Çoruh River was the first project to be completed, in 2005, without resistance. The short-lived campaign by Borçka's residents against construction of the Borçka Dam was quickly contained, with changes made in the project's design and guarantees offered for employment. In the three villages flooded by the Deriner Dam upon completion, local responses did not go beyond murmurs of discontent. Yet, despite the absence of any organized opposition, the completion of all three projects had to be delayed for several years because of their heavy financial burden, and threats and protestations emanating from neighboring countries.[2]

Conceived from the beginning as the costliest project to affect the most densely populated area in the entire valley, the Yusufeli Dam has become subject to even more remarkable vicissitudes of planning and finance. Unlike the other dam projects on the main tributary, the Turkish state sought to overcome the challenge of securing investment for this project by repeatedly signing bilateral agreements with various European and Asian countries to initiate negotiations with their development banks. This determination bore fruit officially by 1997, when a cabinet minister announced on a national TV channel the beginning of construction, as well as the imminent submergence of the town, its seventeen villages, and all their agricultural land. Even if the rumors of a partial submergence had been circulating in the town since the early 1980s, this was the first unequivocal confirmation by a national politician of the massive involuntary displacement and resettlement awaiting the residents. More than a hundred people came together around the Yusufeli Cultural Association (*Yusufeli Kültür Derneği*) to launch a powerful campaign to stop the construction of the project. In the following years, petitions were collected, demonstrations were held, and court cases were brought against various ministries and administrative bodies. Contacts were established with European NGOs to exert pressure on foreign construction companies and export credit agencies (ECAs) involved in the project. As a result, the realization of the project was deferred for more than a decade as two international consortiums were successfully compelled to withdraw one after another. Left with little chance of attracting foreign capital, the Turkish state officially declared in 2010 that it would finance the project itself and started the preparations for the new bidding process. And yet, one could hardly find any trace of the earlier activist energy this time as the majority of residents were now ready to bargain their town's submergence with the state. I therefore ask: why?

This chapter offers an answer by situating the rise and decline of the Yusufeli campaign within the broader changes witnessed in the planning, finance, and construction of large dams in the past three decades. I show that the struggle in Yusufeli essentially succeeded in impeding the construction of the project roughly between 1998 and 2010 by articulating itself into a transnational network of anti-dam/anti-displacement advocacy. Comprised of local communities and activist groups and their allies from the global North, this loose movement managed to counter international capital and its various actors in energy, construction, and financial sectors by resorting to norms and principles formed as a result of previous struggles in areas such as environmental conservation, biological diversity, and human rights. As such, it also contributed to the reshaping of the large dam regime in the 2000s by forcing national governments to seek other means to complete big energy infrastructure projects.

Anthropological literature on resource struggles in general and dam projects in particular are full of debates on the emancipatory potentials and limits of transnational spaces of struggle for local communities and Indigenous groups. For example, whereas Arjun Appadurai (1996: 186) considers the "context-producing activities of the nation-state" as the prison of the Yanomami who were exploited and oppressed by the Brazilian nation-state and makes a plea for post-nationalism as a political strategy, Kaushik Ghosh (2006) by contrast demonstrates the destabilizing effects that strategic "primitivist essentialism" generates on the inclusionary and exclusionary forms of governmentality in the context of the Koel-Karo movement's struggle against the construction of two large dams in India. Another important discussion revolves around whether the ideas and principles of democracy and human rights utilized by local communities against the extraction and appropriation of resources should be attributed to the effects of Western or postcolonial processes and actors, or local mobilization efforts and vernacular creativity (Subramanian 2009).

I make a different point here. The relative strength or weakness of struggles against large infrastructure is subordinated to the differential power of finance. Local campaigns and their allies in the global North could potentially use international norms as leverage against companies and ECAs at a time when national planners' access to direct external funding was essential for implementing construction projects. The peripheral or dependent financialization that emerging capitalist countries have gone through in the early twenty-first century helped to reverse this process as huge capital flows from the capitalist centers coupled with declining interest rates resulted in the appreciation of domestic currencies and the declining of borrowing costs in the global South (Akçay and Güngen 2019). Despite the long-term economic vulnerabilities that it accentuated in southern countries, this form of financial deepening thus paradoxically allowed national governments—in the case of Turkey, the party-state represented by the AKP—to reacquire some of their sovereign capacities previously relegated to transnational actors.

The renationalization of large dam projects thus became a strategy these regimes could afford to employ when attempting to bypass anti-dam/anti-displacement activists' success in blocking flows of international capital. In practice this happened through state funding and the increasing involvement of domestic construction companies, along with local and national politicians. It involved the states' use of tropes as well, such as "energy self-sufficiency" to be able to justify and relativize human suffering and ecological destruction in the name of national interests. The question of why global norms concerning large dam projects failed to influence domestic norms once the borrowing states began to finance the projects themselves (Kadirbeyoğlu 2018), I therefore suggest, should be answered in the light of liquidity that peripheral financialization made available to nation-states from the South. The predatory relations, as well as new aspirations, hopes, and dreams mediated through finance did not only constitute the backbone of authoritarian forms rule in these countries (Mattioli 2020) but also helped to sideline previously successful struggles and campaigns against resource extraction and infrastructural projects. Based on my material on the Yusufeli Dam project, my aim in this chapter is to follow the trajectory by which these struggles came to be rendered ineffective through the detachment of capital from its transnational nexus. To be able to do that, I will first explain under what conditions and in which directions the work of articulation and organization against dam building unfolded in the first place. Since my research began some years after the anti-dam mobilization in the town came to an end, my account of the Yusufeli campaign here exclusively relies on the narratives of local residents and activists, and what little visual and written documentation is left from those years. Yet, this belated arrival was not at all inauspicious, as it allowed me to map out the achievements, tensions, and turning points of the decade-long campaign.

Locals' reflections on and recollections of their struggle and its aftermath recounted here help us to understand some of the circumstances under which construction and finance of large-dams began to be reconfigured around a capital-nation-state assemblage. They also reveal some of the new capacities that the end of the transnational regime of dam building bestowed on the party-state for integrating a community facing dispossession and displacement into its own vision of capitalist development. In this chapter and the following one, I study the spatial and temporal politics of infrastructure to be able to make sense of what these new capacities are and in what ways they are manifested in residents' everyday lives. Here, I begin by describing how the connections between the local and international spaces of activism helped to block energy and construction capital's plans by turning the uncertainties about the dam project and fears about displacement and resettlement into a successful campaign. But I also show that this campaign's success paradoxically contributed to the making of a new regime of capitalist development under which residents' reactions to material changes in their valley could begin to be effectively contained and managed. This shift provides the context in which the anti-dam/anti-displacement struggle in Yusufeli got defeated after nearly a decade. It enables us, as we will see, to understand how some of the former opponents of the project were gradually consigned to applauding the disappearance of their own settlement.

Building Alliances

When the decision to start the construction of the Yusufeli Dam reached the mayor's office at the end of 1997, the reported reply of then mayor Ali Osman Arslan was to tell planners that either the project should be canceled or quickly completed, with the town center being relocated to the village of Öğdem, where he owned large parcels of land. At this early stage, the height of the proposed dam was 74 meters, leading to the submergence of a few villages but not posing a risk for the entire settlement. This initial plan was discarded within months for technical and economic reasons and replaced by a new one that increased the dam's height to 270 meters, the third highest in the world, within the category of arch dams, and the tallest in Turkey. Angered by Ali Osman Bey's reluctance to take action against this new plan that would inundate the entire town center, its seventeen villages, and nearly all the agricultural land in the middle part of the Çoruh Valley, a group of shop owners and local politicians from different political parties met with a lawyer at the beginning of 1998, founding the Yusufeli Cultural Association.

The period between 1998 and 2007 is remembered in particular by Yusufeli residents as a remarkably vibrant time, with the majority of them taking part in political meetings and demonstrations for the first time in their lives. Led by the lawyer Ragıp Bey and Muhsin Abi, the local association organized meetings with village heads and invited academics to give presentations about the scale of destruction and displacement awaiting local residents. In the early 2000s, demonstrations were held in the town center to pressure the state either to go back to the earlier 74 meters plan or to cancel the project completely. A blurry video recording from this time uploaded to YouTube shows hundreds of men, including heads of various local party branches, marching along the main street chanting "We are not against the dam but getting drowned!" (*Baraja değil, boğulmaya karşıyız!*),[3] reiterating the call for implementing the first plan with 74 meters. Occasionally, even local bullfighting festivals and Bayram celebrations were turned into political arenas where residents protested regional governors or collectively swam in the cold Çoruh River to show that their fate was intimately bound to its waters. More importantly, in 2002 the local association filed suit against the Ministry of Energy and the State Hydraulic Works (DSI), demanding the project be canceled.

In addition to organizing protests in Yusufeli, local activists also played an important role in lobbying against the project by activating social and political networks formed over the years as a consequence of many decades of migration from the region to big cities. After making contact with MPs and prominent bureaucrats hailing from Yusufeli and other towns in the vicinity of Artvin, in 2003 association members visited President Süleyman Demirel and officials from the DSI, presenting a petition opposing the dam project signed by almost three thousand residents and voicing their objection to Yusufeli's submergence.

Co-locality organizations (*hemşeri dernekleri*) also took part in the campaign, prompted in particular by the devastating earthquake that struck the Marmara region in 1999, killing forty thousand people. Up until then, for first- and second-generation migrants—many of whom still owned a family home and small parcels of land—Yusufeli primarily served as a site where they spent low-cost summer vacations swimming in the river and relaxing in coffeehouses. The massive destruction caused in the centers of Istanbul and Izmit, along with repeated warnings voiced by geologists regarding the increasing possibility of an earthquake hitting the region again, made Yusufeli natives view their hometown as a potential refuge in case of emergency. Several meetings the association organized with co-locality groups in 2002 and 2003 fueled anti-dam sentiments, helping to make the campaign more visible nationally. By all accounts, it also strengthened the residents' own prospects for defeating the state and saving the town from flooding.

While the Yusufeli association to a large degree succeeded in translating widespread fears and uncertainties about the submergence of the town into a unified struggle, this achievement was not without new challenges. Some Yusufeli shop owners publicly announced their support for the construction of the dam in the early 2000s, based on expectations that an influx of migrant workers and engineers would boost business. They even ran a rival local association for a short period, which sought to arrange deals with prospective construction companies, obligating them to purchase goods from association members. Even though the initial consensus to build a campaign beyond political differences and party affiliations remained intact, as exemplified by the collegial relations Zeliha maintained with ultra-nationalist anti-dam activists, the election of the MHP's candidate, Yusuf Sağlam, as mayor in 1999—based on his anti-dam credentials and with the support of the local association—estranged some association members and voters from other political parties. Also by the early 2000s, locals possessing more than the average two or three *dönüm* (a thousand square meters) of agricultural land began to criticize local association members for their efforts in an online discussion forum to block the dam project, although this was not limited exclusively to the wealthier townspeople. While some residents were motivated by the compensation payments they would receive after their property was expropriated, others anticipated, not unlike the former mayor, that their home villages might become the center of the new settlement area.

Uncertainty around the location of a new town center began to create deeper tensions and divisions among other residents as well. Surveys conducted by the DSI to gauge public opinion added to the confusion, as several new names began to circulate simultaneously as possible candidates. While some tried to convince fellow villagers their settlement should host the new Yusufeli, those living in and around the town center vehemently objected, insisting that a place close to its current location, which sits more or less at the heart of the sixty-three villages of Yusufeli, should be chosen. It was not until 2008 that this issue was officially settled, when a nearby hilltop known as Yansıtıcılar or Sakut'un Deresi (Sakut's Stream) was announced as the new settlement area in legislation passed by parliament. Until then, and even later, disagreement over the location of the new settlement area created distractions from the campaign and gradually bestowed a sense of inevitability on the project to normalize the idea that land should be submerged for energy production. I will explore this point more in detail in the next chapter.

A public meeting organized by the cultural association in the summer of 2005 became a fiasco, as participation remained limited, according to activists, due to very hot weather and technical problems, but it can also be seen as an early sign of decline afflicting the energies behind the campaign. At this point Ragıp Bey and Muhsin Abi decided to concentrate on the legal challenge rather than organize more events and protests locally. They also believed dissenting voices in the town would remain ineffectual in altering the majority of the local population's objection to the project as long as the fight led by Zeliha and her friends in the international arena continued to prevent the start of construction.

Having migrated to Germany from Yusufeli in her teens, Zeliha ascribes the beginnings of her involvement in the Yusufeli campaign to a demonstration organized near Erfurt at the end of the 1990s against another notorious mammoth project: Belo Monte Dam in the Brazilian Amazon. Touched deeply by the plight of the Indigenous communities threatened with displacement by the completion of the project and inspired by this show of solidarity thousands of kilometers away from the Amazon, she got in touch with the activists from Yusufeli in the early 2000s. Accompanied by Ahsen, another left-oriented woman based in Berlin, and working closely with Erhan, a Kurdish-German activist campaigning against the construction of the Ilısu Dam project that would submerge the historical town Hasankeyf in the eastern part of Turkey, she played a crucial role in bringing the local campaign in close contact with transnational activist networks.

In our only face-to-face conversation in Berlin in 2013, Zeliha shed light on the lengthy process she and her activist friends used to exert pressure on European companies and ECAs to withdraw from the Yusufeli Dam project. As she explained to me:

> There exists a long process before ECAs provide credits to European companies. Locals' consent is important. The ecological impact of the project also needs to be taken into consideration. International standards have to be addressed when it comes to environmental protection and resettlement. Equator Principles[4] and the EU criteria were important for Çoruh. So, our first aim was to stop the export loan by putting pressure on European governments and companies. . . . First, we get information [about the companies and the ECAs] from [European] NGOs. We write letters to the managers of the company, saying that the people from Yusufeli have no clue about the project's details, the project does not meet the international standards. . . . This way, we managed to oust two consortiums; first one led by the British company Amec, and then one comprising the Swiss company Alstom, the French company Coyne & Bellier and their domestic subsidiary Doğuş and its construction company Dolsar.

The Yusufeli campaign's success in impeding the project essentially relied on its ability to internationalize its own tactics, mirroring the transnational composition of large dam finance and planning. In the aftermath of the Narmada Valley struggle in India, such tactics had become well understood, and put to use by other anti-dam/anti-displacement groups throughout the world. After the success of the Save the Narmada Movement's (Narmada Bachao Andolan) inspiring campaign against the construction of several large dams in the Narmada River Valley, in the early 1990s the World Bank, the main financier of the controversial Sardar Sarovar Dam project, set up an independent committee called the World Commission on Dams (WCD) to assess in detail the social and environmental effects of large dams around the world (Routledge 2003). Published in 2000, the final report by the WCD contained much criticism of national dam planners, transnational construction and energy companies, and the Bank itself, and proposed a series of recommendations to "stakeholders" to address the social and environmental problems associated with large dams. By almost all accounts, the WCD's initial goal to develop and promote international standards for dam building was a failure, with nation-states such as China, India, and Turkey rejecting its recommendations, while the World Bank and a majority of development banks abstained from fully adopting the report's proposals (Fujikura and Nakayama 2009). Regardless of its original intentions and final outcomes, the process that started with the establishment of the WCD was instrumental in opening a new battleground between proponents and opponents of large dam projects.

A few factors may have placed Zeliha, Alisen, and their allies in European NGOs such as Les Amis de la Terre (France), The Corner House (UK), and Erklärung von Bern (now known as Public Eye, from Switzerland) in an advantageous position from which to oppose the consortiums that sought to take part in the Yusufeli project. First, it was not until 2007 that an environmental impact assessment (EIA) report was prepared for the Yusufeli Dam project, because of a Turkish law stipulating that projects developed before 1993 are exempt from filing such assessments.[5] Some observers have argued that EIAs had become one essential criterion for receiving international funding following the WCD report's publication (Sneddon and Fox 2008). The absence of an EIA report was repeatedly raised by activists when communicating with officials from international consortiums and ECAs. They also stressed this point when, together with activists from the Hasankeyf campaign, they met with representatives of various European political parties in Brussels in the mid-2000s. COFACE, the French ECA approached by the second construction consortium, explicitly named Turkey's refusal to compile an EIA report as its reason for withdrawing from the project. But even after the Turkish state bowed to international pressure, giving the green light for preparing a report by a company called ENCON[6] in 2007 (which itself was severely criticized by the local association for its ties to the Doğuş Company), it was still not enough to secure export credit insurance. Swiss Export Risk Insurance (SERV), which got involved in the project after 2004, also during the time the second international consortium was trying to begin construction, was one of only a handful of banks to have officially adopted the WCD report's recommendations. In 2004, shortly before negotiations over Turkey's full membership in the European Union officially began, the EU also addressed the guidelines set by the report in a directive requiring member states to respect these standards during large dam projects' development phase (Fujikura and Nakayama, 2009). The notable lack of concrete plans for resettlement and environmental protection contradicted the criteria specified in these guidelines.

Even if these objections may not have decisively deterred the consortiums and the ECAs, what they clearly achieved was to render the Yusufeli project riskier and more uncertain from the point of view of capital. As Zeliha put it: "The companies withdrew not because they were afraid of us. They left because the delays and interruptions that we caused in the project negatively affected their [estimated] rate of return." Her remark indicates how the main tactic of transnational anti-dam/anti-displacement activism has involved defying capital by tampering with the temporality of its flows and accumulation. By slowing down an already lengthy capital investment horizon, where returns commenced only after dam construction was complete, the reservoir filled, and electricity production began, they gave international companies no possibility other than to withdraw from the project.

A meeting organized at a hotel in Yusufeli in October 2007 with officials from the DSI, delegates of the Swiss and French ECAs and Doğuş Company, members of the Yusufeli Cultural Association, and an employee from Erklärung von Bern, the aforementioned Swiss NGO, specialized in assessing the environmental and human rights aspects of international projects run or funded by Swiss companies and banks, laid bare how the cross-border activist network could work powerfully against the international regime of dam building on the ground. Over the course of the two days of talks, while some local residents who had previously worked in Europe as "guest workers" challenged consortium officials in flawless French, activists from the campaign repeatedly argued that the project failed to meet international norms and standards on environmental protection and resettlement. Interviews with local residents conducted by the Swiss NGO's employee during the talks revealed the dearth of information residents had received concerning the project, including where and when the new settlement area would be built. In the months after this meeting, she compiled a damning report that concluded:

> DSI has no interest in resettling the affected people. The false claims by the DSI that people would rather take cash compensation shows that the DSI aims to expropriate people and abandon them to their fate in order to save money and a lot of trouble, e.g., by not having to build houses in rocky steep hills and create gardens where nothing can grow. . . . Therefore an approval of the project would imply a forced displacement of the affected people which seriously contradicts the standards of the COFACE and the SERV, as well as the World Bank. (Eberlein 2007)[7]

SERV was forced to decide against providing credits for the dam, and at the end of 2007 the second consortium officially withdrew from the project. Around the same time, the legal case brought against the Ministry of Energy and the DSI by the local association concluded. Estimated by local activists to result in a loss of around two billion dollars if completed, the dam project was halted based on concerns over its economic viability. However, following the Ministry's appeal this decision was quashed and the case sent back to the lower court for review. While reiterating his skepticism of the project's economic viability, this time the judge gave the green light for construction to begin, reportedly under extreme pressure from Ministry officials. While the Turkish state continued to search for new international donors to no avail, in early 2009 the design of the project was abruptly changed from a rock-filled to a cement-filled arch 270 meters in height. Shortly after the announcement by the government that the project would now be financed exclusively by the Turkish state, a new auction took place in 2012 to deliver the project to a consortium formed by three leading domestic construction companies: LİMAK, KOLİN, and Cengiz. Yusufeli Dam would now be constructed through a public-private partnership and then run by the state upon completion.

On 26 February 2013, the foundation stone of the Yusufeli Dam was laid in an ostentatious ceremony attended by cabinet ministers and high-ranking bureaucrats, and inaugurated by a satellite link with then Prime Minister Recep Tayyip Erdoğan. Only two months before, the same coterie of politicians and state officials had attended the opening of the Deriner Dam, another hydropower facility in the region and the tallest completed thus far in Turkey, with a height of 249 meters. Thousands of people from Yusufeli and neighboring towns and villages enthusiastically filled the town center that day in support of the prime minister and his government's ambitious hydroenergy projects. According to Muhsin Abi however, they were merely applauding the future destruction of their land and livelihoods.

After the Transnational

Anti-dam/anti-displacement activism altered the political economy of development by targeting the composition of dam finance and planning (Khagram 2004). Through transnational coordination and action, these activists successfully helped to establish global norms and standards, which various campaigns used to hinder and eventually unravel the international large dam regime. First established in the 1950s, this regime continued to expand over the course of the following four decades, with the addition of new construction companies, development banks, and ECAs. Out of its defeat emerged a new regime impervious to the tactics and tools of cross-border struggle. In that sense, renationalization of the Yusufeli Dam project severely curbed the strength of the campaign, stripping activists of valuable instruments such as guidelines and standards concerning environment and resettlement. Some of the regulations through which credit agencies and transnational consortiums could previously be made accountable (if only in part) lost their binding character against the project's new financier, thus weakening the struggle against the construction of the Yusufeli Dam project.

This fate was shared in one way or another in the case of all three other large-dam projects mentioned earlier. Sardar Sarovar Dam project was completed with funding from the Indian government after the World Bank was forced to cancel the remainder of its loan. Direct action protests staged by activists and links formed with international NGOs and advocacy groups had been crucial in compelling the Bank and the Indian Supreme Court to admit that criteria laid out for displacement, resettlement, and rehabilitation had not been met by project planners (Kadirbeyoğlu 2018). Notwithstanding the internal conflicts accumulated over decades which undermined the campaign, the nationalization of the project rendered Save the Narmada campaign to a large extent powerless against the judicial and political maneuvering of subsequent Indian governments, who continued to raise the height of the dam to submerge even more areas with impunity (Chakravarti 2017). Transnational activism tactics also lost their political influence in the case of the Belo Monte project after the role of external finance diminished. Under the Worker's Party government, after 2008 the Brazilian state, while acknowledging the importance of environmental and human rights norms, subsequently diluted them by adopting a "green development" strategy, using party loyalty to create divisions among the campaigners (Bratman 2014). The campaign against the Ilısu Dam project continued to rely on its international connections even after European banks were compelled to suspend funding, thanks to the efforts of the Kurdish diaspora living abroad and widespread concern over the submergence of the ancient town of Hasankeyf. But still it lost momentum, with a majority of local residents gradually withdrawing from the campaign after the AKP government used both economic incentives and political pressure to break opposition to the project (İlhan 2016).[8]

In an important body of literature, ethnographers and political theorists have recently returned to the question of how capital rearranges the relations of space to sovereign power in the context of contemporary processes of resource extraction and appropriation. While a particular investment in and figuration of space as territorial control holds a central place in classical understandings of state sovereignty, by contrast these accounts are concerned with how commercial powers generate territories where sovereign rule is not entirely invested in the state form. "Enclaves of extraction" (Ferguson 2005) and "economic corridors" (Dey and Grappi 2015), to name two recent examples, embody territorialities where capital bypasses nation-state frames as it "hops" from one place to another, or where sovereign capacities are reduced to one out of many other elements facilitating capital's operations of extraction, financialization, and logistics (Mezzadra and Neilson 2019). At the same time, what has been described as "zones of imperial extraction" (Gordillo 2019)—for example, the soybean production that connects rural Argentine to other locations world-wide through supply chains and logistical infrastructures—reveal the planetary density of networks of value capture under which nation-states and imperialistic forces counter and coalesce the power of global capital.[9]

Even though I find these observations accurate and valuable, my material on Turkey points to an opposite tendency: economies of construction and destruction become a means through which state power and capacity is expanded and reinforced. I do not mean to suggest that global capital and expertise are entirely absent in large infrastructure projects. Rather, I suggest the void left by the diminished role of transnational consortiums and banks in the global South has increasingly been filled by national governments, bringing processes of accumulation—especially in extractive and infrastructure industries—under the control of political parties and their leaders. In addition to the success of transnational forms of activism in the context of large dams, I also recognize the availability of global liquidity as the background condition of this new regime of infrastructural development, made possible by the sinking of interest rates in the global North over the course of the first two decades of the twenty-first century (Madra and Özselçuk 2019). During this period, the appreciation of national currencies, swelling of central bank reserves, and a steady increase in growth rates enabled countries like Turkey to plan or complete large construction projects entirely through state funding. Notwithstanding the fact that needed funding may have been acquired from international financial markets through external borrowing, large construction projects nonetheless became "national projects." Not only did domestic companies replace international actors but also the politicians from the ruling AKP began to take critical decisions, sometimes in coordination with, other times against the wishes of administrative bodies such as the DSI, about the plans for construction, expropriation, and resettlement.

One important aspect of this shift concerns the increasing role sovereign decision-making plays in shaping the economy. Under the leadership of Recep Tayyip Erdoğan, first as prime minister and then as president, the AKP awarded contracts for construction and extraction projects to a handful of domestic companies who staunchly backed the party. In a recent report (2018) and the accompanying data base released by the World Bank, the companies that form the domestic consortium behind the Yusufeli Dam project after 2012—LİMAK, KOLİN, and Cengiz—are listed, together with two other Turkish corporations, among the top ten companies worldwide with the highest number of tenders won for public infrastructure projects between 1990 and 2018. In addition to dam projects, these companies were involved in several other flagship projects, including airports, highways, and bridges. Overall, Turkey placed fourth, after Brazil, India, and China, in the amount of private investment in infrastructure during the same period. While these public-private partnerships signal an alliance between domestic capitalists and the state that drives processes of dispossession under neoliberalism (Levien 2018), arguably they are revelatory of an even more ambitious reconfiguration of state-capital relations.

Especially after Turkey's construction-based growth strategy began to show signs of "hitting the wall" (Adaman et al. 2014) after 2013, the AKP made a number of significant interventions to buoy the economy. In addition to restructuring the debt of the aforementioned companies under the Department of the Treasury guarantee scheme, the state also offered billions of dollars a year in loans as a feature of some construction projects. If they continue to endure financial stress, these companies may obtain even more loans in the next years from the recently established Wealth Fund.[10]

This chapter argued that the detachment of dam capital from its transnational nexus became the background condition by which struggles against large infrastructural projects lost their momentum. The availability of global liquidity made possible by the conditions of the global crisis set in motion the dependent financialization of the countries from the global South. As a result, large dams such as the Yusufeli Dam project could be renationalized or completed without resort to direct foreign investment. The entrenchment of national governments and their leaders' sovereign capacities and powers is one of the significant consequences of this development that I explored here.

But I also showed that resistance contributed to reconfiguring capitalist development under a new set of conditions, against which it was, paradoxically, rendered ineffective. It was, after all, the anti-dam/anti-displacement campaigns, together with their allies from the global North, that achieved blocking the flows of international dam capital by resorting to global norms related to resettlement and environmental protection. My argument regarding the interruption of politics resulting from a shift from transnational to national regimes of dam finance and construction in that sense invokes an idea advanced by theorists coming from the Italian workerist/post-workerist tradition: conflict is an internal mechanism of capitalist development and therefore capital accumulation is contingent on struggles (Tronti 2019). Stated differently, this means that resistance precedes domination by capital, but is parasitically contained by the latter for its survival (Negri 2005, 2008). Whereas workerist and post-workerist writers were mainly concerned with labor's struggle against capital however, anti-dam/anti-displacement campaigns around the world echoed in their fights what Nancy Fraser (2014, 2016) calls "boundary struggles": a form of conflict around "capitalism's institutional divisions" which separate "economy from polity, production from reproduction, human from non-human nature" (2014: 68). In that sense, anti-dam/displacement activism cannot be explained by a single economic or political cause or agent. It should instead be seen as bringing together a multitude of subjects mobilizing around grievances and injuries related to a wide array of issues including class, gender, identity, and concern for the environment.

Beyond the intricate web of political and financial connections that underlies the symbiotic relation between the party-state and certain factions of capital that it helped to strengthen in the first place, bulldozer capitalism's reconstitution around a capital-nation-state assemblage produced another important effect that I now turn to: steering of the residents' experiences and sensibilities of time and space at the local level.

Notes

1. While I referred to the "Çoruh Basin Development Plan" in my earlier publications (Evren 2016), the official title is the "Çoruh Energy Plan." Once again, this confusion stems from the fact that most politicians and bureaucrats use both names interchangeably in conversations.
2. The Çoruh River originates from the northwest of the Erzurum-Kars Plateau and empties into the Black Sea near Batum, Georgia. Approximately ninety-one percent of the river lies on the Turkish side of the border. Anecdotal information shared with me by state bureaucrats and hydrologists indicates that the USSR's objection to the Plan, until the Soviet Union's demise in 1991, was the main reason behind the project's long delay. This objection also informed the planners' rather unusual decision to begin the construction from downstream with the least costly Muratlı Dam project in 1991, months after the fall of the USSR. After its independence, the Georgian state voiced similar concerns about the Plan, with less success. For more on the political implications of the Çoruh River's transboundary nature and its impact on Turkish-Georgian relations, see Klaphake and Scheumann (2011).
3. For the video of this protest march in Yusufeli, see www.youtube.com/watch?v=lIj_IEVeGJQ.
4. Equator Principles refer to a series of environmental and social guidelines developed by leading private banks based on the standards of the World Bank and its International Finance Corporations (IFC) to assess the projects that they are asked to fund (Amalric 2005). However, the lack of any enforcement mechanisms until 2012 rendered the practical implementation of these standards difficult. After the 2012 edition of IFC's Sustainability Framework, all project funding applications were obliged to go through an initial credit review process to determine how the clients manage the social and environmental risks of their projects.
5. 1993 was set as the threshold for applying impact assessment in environmental disputes in the first legislation ever passed in the parliament on this issue on the same year. Apparently, this was done by the ruling coalition government to circumvent the growing campaign against the construction of the third bridge on the Bosporus, then planned to pass through the old neighborhood of Arnavutköy on the European side of Istanbul (personal communication with Sinan Erensü). Similar to the Yusufeli

Dam project, there would be no requirement to compile an EIA report for this project as it had been planned years ago. Redesigned under the AKP governments, the third bridge or the Yavuz Sultan Selim Bridge, as it was later named, was eventually completed in 2016.

6. Actually, Doğuş Company asked ENCON to prepare an EIA report as early as 1998 but its publication was canceled at the last minute by the company. Almost twenty years later, Arda Bey, a wealthy and well-connected resident from Yusufeli, discovered this report and posted parts of it on his Facebook page. The projection made by ENCON that Yusufeli's population will severely decrease after the resettlement, may have dissuaded Doğuş Company to release its results as this fact would be clearly in the local association's advantage. Despite my repeated requests, Arda Bey refused to share the entire report with me.
7. For an insightful academic article that the same employee from the Erklärung von Bern later wrote together with other NGO workers and activists on the role ECAs play in dam development and anti-dam/anti-displacement campaigns, see Eberlein et al. (2010).
8. Despite the protestations by activists and European NGOs, an Austrian company eventually supplied the turbines for the power plant and provided technical support for the project. Ilısu Dam began to fill its reservoir in July 2019 and shortly after most parts of Hasankeyf were flooded. Since 2009, environmentalists involved in the Ilısu campaign have been building regional connections, including with the Bedouin community affected by the project on the other side of the border and the activists from Nature Iraq (İlhan 2016).
9. For two other recent works on Latin America, which emphasize on the planetary scale that networks of logistics and supply chains have reached in recent decades, see Hetherington (2020) and Arboleda (2020).
10. Founded in August 2016, the Turkey Wealth Fund (*Türkiye Varlık Fonu*) owns the shares of several state enterprises banks and companies active in telecommunication, mining, aviation, and agriculture.

— Chapter 2 —

Politics of Time and Space

Looking Forward

After the withdrawal of the last transnational consortium at the end of 2007, residents quietly continued to follow rumors about the dam's future for the next few years. In the meantime, the AKP was steadily consolidating its power, swallowing the constituencies of once powerful center-right parties and thereby transforming the political map of small towns and cities across Turkey. Local businessmen, lawyers, teachers, and shop owners from Yusufeli joined the party around this time, making possible a historic AKP victory in the 2009 municipal elections. "From that point onwards," Muhsin Abi explained, "it was all about whether you [were] pro- or anti-AKP, and our campaign against the dam suffered badly from that."

Several people who talked to me in 2013 still spoke critically of the Yusufeli Dam project and mentioned the economic and social difficulties that awaited them upon completion of the project. Some of them pointed to the barren hilltop towering over the town center, officially chosen as the new settlement area, to wonder what kind of life would await them in the new Yusufeli. Others lamented the fact that they were about to lose not only their homes and streets but also memories, habitual ways of living, culture, and identity. Remarkably, however, the same interlocutors often explicitly put the blame for their grievances on the decade-long anti-displacement struggle and its main protagonists as much as on the dam project itself. Having succeeded in delaying the project twice but having fallen short of achieving its cancellation, the local activists were accused in these narratives of prolonging the jeopardized existence in the town, condemning an entire generation of people to continue their lives in a state of limbo. After dismissing my questions about the three remaining activists' plans to file a new lawsuit against the project in summer 2013, the imam of a local mosque and a longtime resident, who had witnessed the ebbs and flows of the project since early 1980s, summarized the dominant affect at this point as follows: "We feel alienated. We feel fatigued. In all these years, we couldn't plan our future because of the dam project and the interruptions to its realization. We want to plan our future now, we want to be able to look forward (*önümüzü görmek istiyoruz*)."

Ruination, if we follow Ann Laura Stoler's (2008: 195) description, may alternatively refer to "an act perpetrated, a condition to which one is subject, and a cause of loss." The protracted devastation along the Çoruh Valley by other dam constructions coincided with the repeated deferral of the Yusufeli Dam project to entail all three definitions. As a condition of suffering that local people such as my imam friend recognized, as alienation and fatigue stemming from prolonged uncertainties, ruination resonates with a peculiar version of a kind of injury that Rob Nixon (2013) calls "slow violence": the suspended, slow-motion, and out-of-sight devastation experienced by the target communities not only of mega-dams but also of atomic tests and oil spills. What makes slow violence peculiar in the context at hand is that, whereas the suffering it has elicited served in part to sustain an oppositional political formation for almost a decade, the majority now sees this suffering as the outcome of the political agency behind this formation.

This chapter examines how the experiences of time and space at the local level come to play a role in the politics of capitalist development. Like the regional head of DSI in Artvin who once proudly described to me the mega-dam projects on the Çoruh River as "jewel necklaces" (*gerdanlık*), most planners of large-scale infrastructures embody a high-modernist impulse necessarily indifferent to local histories, attachments, and lived experiences. Target communities' reactions to spatial and temporal changes caused by the transformations of the built environment and ways of living, however, continue to exert an influence on their political responses. Here, I try to understand under what conditions "looking forward" came to gradually replace the fears about destruction and dispossession, and the uncertainties about displacement and resettlement. The material expectations from and investments in a future to come that the party-state skillfully controlled, I argue, turned into the ground on which the activist energies directed against the Yusufeli Dam became exhausted.

Few studies on Turkey have so far explicitly dealt with how citizens' experiences of time and space mediate their relations with the state. Leila Harris's (2009) work on the arrival of canalet irrigation to the marginal rural space of the Harran plain made the point that state-society relations are reshaped partly horizontally through the creation of new spaces and life practices, which buttress the state's attempts to forge a modern Turkish future. Anna Secor (2007), on the other hand, looked at the relational production of the citizen and the state with an emphasis on spatiotemporal techniques and the affects, such as guilt and despair, which these techniques elicit among lower class men and women in Istanbul. Finally, Alize Arıcan's (2020) important ethnography of one of the first state-led urban transformation projects in Istanbul read the delays and suspensions in construction as generative of political and economic asymmetries between workers, subcontractors, and AKP politicians, pointing to an understanding of accumulation and dispossession that goes beyond the linear temporalities that capitalism imposes.

This chapter intends to contribute to this literature in two ways: first, it moves the discussion away from the practices of a generic (developmental) state to flesh out in what ways the new conflations between party networks and sovereign decision-making structures, put into use across national and local scales, help to manage and steer people's desires and anticipations. My material highlights that following the AKP's rise to power, first nationwide and then at the municipal level, its local branch, prominent members with personal ties to the town, and even President Erdoğan himself in some cases, intervened to address some of the major concerns while making promises for resolving others. Second, it shows that the party-state's active encouragement of residents to participate in the negotiation of their expectations helps to establish an intersubjective relation reminiscent of what Lauren Berlant (2011) calls "cruel optimism": a relation by which the desires that one attaches to something hinder his or her flourishing. Below and in the following chapters, I think of such attachments, and their spatial and temporal dimensions constitutive of a hegemonic project that the AKP has actively built in the past decade and a half. But I also point to those instances and events that reveal the limits of this project.

By studying "looking forward" as the key future orientation that the party-state helps to construct and colonize, my aim is to contribute to the study of class dynamics underlying the rise and success of the AKP regime. I highlight in this chapter and the following ones that local businessmen, in particular those who become active in the construction business and/or personally enter the local party networks, act as the umbilical cord between the party-state and ordinary citizens, channeling some of the latter's expectations related to compensation, resettlement, employment, and real estate.

Accommodating Ruination

Around the time that Zeliha and other activists were making use of the transnational space of mobilization to counter the international regime of dam finance and planning, the local time-space of Yusufeli was going through enormous changes. The belated construction of the first three dams quickly set in motion the drastic deterioration of the physical landscape in the Çoruh Valley. Billows of dust rising from construction sites after late-night controlled explosions would be blamed for causing decreases in agricultural yields, just like the changes in the Mediterranean microclimate of the valley would be attributed to the drastic damming of the river. With every plot of land expropriated and bulldozed to make space for building sites and sand pits, and with every landmark, such as a historical fountain or bridge, that simultaneously vanished, material destruction created an ever-growing sense of loss in the valley's residents.

From the late 1990s onwards, new road and tunnel building projects across the valley began to severely interrupt transportation to and from Yusufeli. Some houses, several plots of agricultural land, and, in one case, an entire village close to a construction site, were quickly expropriated in order to open space. "It was getting closer," one man from Yusufeli would tell me, describing his worries about this drastic transformation in the region. By the mid-2000s, sudden decreases in the flow and volume of the Çoruh River began to severely impact commercial rafting activities in the area, adding an economic dimension to locals' perturbation. Throughout this period state agencies exacerbated this material devastation by dissuading the town's residents from repairing or rebuilding their crumbling houses or local infrastructure, telling them repeatedly to expect the beginning of dam construction and their attendant resettlement within a few years. Like the "new urban zone" of Thủ Thiêm in Vietnam, where the proliferation of strangers—including planners, journalists, urban studies scholars, and heroin addicts—complicated the local sense of space and identity (Harms 2013), the arrival in Yusufeli of engineers, company representatives, and construction workers introduced an element of "strangeness" to the formerly close-knit community. Waiting for the completion of damming on the Çoruh River, in short, introduced an immense sense of uncertainty into the lives of this community's members.

Radical forms of uncertainty that people encounter as they wait for personal or societal transformation have received attention in recent years among anthropologists and other social scientists. In particular, they asked how waiting is connected to the social worlds in which we live (Hage 2009) and in what ways its rhythms and routines serve to consolidate, produce, or transform political subjectivities (Biner and Biner 2019). For example, Javier Auyero has extensively written about how stalled time leads to the production of compliant citizens through the dissemination of a sense of uncertainty in the cases of an environmental catastrophe (Auyero and Swistun 2009) and a bureaucratic setting (Auyero 2011). Erik Harms (2013), on the other hand, differentiated between oppressive and economically productive modalities of waiting, showing that whereas endless waiting can appear as a mechanism of control by leading to the absorption of residents' political energy, waiting in the form of "hanging out" can help to build up connections that can be translated into monetary means. The future-oriented temporality of planning and development no doubt already left its mark on Yusufeli and its people, as representatives of Doğuş Company, which was the domestic subsidiary of the second international consortium, and state officials reiterated promises of employment, high compensation payments, and a new and a better settlement area for locals. And yet, opposition to the dam project was also fueled by fears of social, moral, and physical destruction and disintegration. In addition to using the international space of dam planning and finance, activists from the Yusufeli campaign could block the dam construction also by making visible residents' anxieties about displacement and resettlement. Their cross-party efforts served, for a certain period of time, to confine pro-dam voices and the political effects of their future-oriented monetary expectations.

Following its renationalization, party members and state officials increasingly strived to present the Yusufeli Dam as an AKP project made possible by its leader. Reportedly, during the opening ceremony in the town in 2013, they even offered to rename it as the "Recep Tayyip Erdoğan Dam." Even though this proposal was outright rejected by Erdoğan himself, such material and symbolic maneuvers by the party-state, and its local members and sympathizers, aggravated the divisions among the residents by framing the opposition to the project as an expression of hostility against the party and its leader itself.

Even ordinary ways of reflecting on the transformations of the valley and the town began to be closely followed and censured. Left-leaning friends teaching at the local high school complained that pro-AKP teachers disapproved, and in some cases, tried to prevent cultural activities such as drama or music performances that questioned the destructive effects of dam projects. An issue of the local high school magazine, which brought together poems, texts, and drawings by students that mourn the ongoing devastation of the valley and the coming submergence of the town and its villages, was severely criticized by the school principal. Its circulation in the town was even temporarily halted. One teacher friend nevertheless went ahead and staged an entertaining play he wrote in April 2014 about three villagers who decide to establish the "New Yusufeli" in a deserted island, subtly poking fun at the possible forms of estrangement awaiting the locals in the new settlement area.

Displacement and resettlement were at the same time normalized not simply as a price that the residents have to pay for the production of electricity, which is deemed necessary for the country to achieve energy self-sufficiency, but also as an unfortunate predicament that the valley's previous residents had encountered and successfully overcome. Repeating a point made by Erdoğan himself during the ceremony in February 2013, Eyüp Bey, the mayor of Yusufeli after 2009, and other figures from the party often stressed in their interviews with journalists and in articles they published in the municipality's in-house journal that the town center's location had previously been changed six times.[1] This new discursive strategy eventually gave inspiration to a motto, "Yusufeli is searching for its seventh location" (*Yusufeli Yedinci Yerini Arıyor*), which regularly appeared as the headline of news reports published in the local and national media. In 2017, it was also used as the title of a booklet published by the mayor's office, which shares the results of a survey conducted in 2012 on the residents' views on different aspects of the dam project. According to this study, more than half of all residents continue to oppose the Yusufeli Dam project and the submergence of the town.[2] The mayor's introduction, as well as the texts posted on the municipal website to promote the booklet, however, presented the project and its effects as inevitable.

This was indeed the message Eyüp Bey and other members of the local AKP branch repeatedly conveyed to their fellow residents after the new consortium took over: it is too late to stop the project, but we can negotiate our demands and therefore minimize our grievances only if we come together to support the AKP. This message found widespread support among Yusufeli's inhabitants who felt exhausted by waiting in uncertainty for decades, not being able to tell if their settlement would remain intact or be submerged by the dam's reservoir within a couple of years, all while enduring in their daily lives the destructive effects of other construction projects. Earlier, the protracted ruination of the valley might have helped to intensify everyday forms of discontent towards the Yusufeli project and the planned displacement. In the course of the 2010s, however, it set the background against which negotiations with the state over a wide range of demands, as well as those visions for speculative accumulation and commodification (as we will see in Chapter 4) took shape.

Incitements to Negotiation

Yet, it was prominent figures from the party, with personal connections to the town or the region, which became instrumental in inculcating among residents the idea that if it is negotiated, submergence and displacement may not necessarily lead to dispossession as the effects of destruction can be made monetarily rewarding. I observed firsthand the excitement and frustration encounters with these politicians produced when Kadir Topbaş, the then mayor of Istanbul, accompanied by a group of businessmen, arrived for a short visit in the summer of 2014. The hours of preparation that had gone into setting up a sound system and decorating nearby buildings and trees with balloons and banners bearing AKP logos initially seemed justified when he stepped out of the bus to salute the couple of hundred locals impatiently waiting for him along the main street. A table set for him featured locally grown grapes, presumably from one of the last vineyards to have survived the ongoing destruction of the valley, and a local *kemençe* player provided a musical fanfare. Speaking into a creaky microphone, Topbaş thanked the residents and mumbled for a couple of minutes, praising the taste of the grapes. He ended his speech with the cryptic remark, "We should take our past, our culture into the future," then quickly left with his entourage to spend the night at his family house in Barhal, to the disappointment of his fellow townspeople. "He left without really saying anything," was one complaint I heard among the crowd before heading to my hotel. Nonetheless, a common past and place of origin that the residents imagined to share with him and former cabinet ministers from the region such as Faruk Çelik, fostered trust in the present and built up positive expectations for the future. Bulldozer capitalism begins with claiming the river as free raw material before planning the conversion of its motions into electricity through large-scale infrastructure. Its effectiveness, however, seems to be ensured by capturing and putting into use more symbolic forms of commons built on an assumed identity. Imagined or real, a shared history, identity, and worldview that I qualify here as conservative, therefore comes to electrify the attachments to these political figures, inciting people to envisage bargaining the project's effects of expropriation, displacement, and resettlement. Conservative commons, in other words, ensures the production of capitalist value within a framework of consensual political practice.

I found that "looking forward" has indeed emerged as the emblematic gesture at this juncture. This expression primarily designates locals' orientations to a future in which they will have foreseen and successfully survived the worst effects of their valley's and town's submergence for electricity production. At the same time, this phrase also eloquently captures the anticipatory logics and calculations that underlie the economic desires of the same residents. Some version of a monetary favor bestowed on the residents loomed large among the countless ideas and visions that they thought these politicians could help realize. Retired bank employee Kemal Bey invoked the technical term *şerefiye parası* (goodwill compensation) to suggest that the calculations of the value of property which will be expropriated by the state should become subject to a special assessment that would be reflected in an increase in compensation payments. Kaan Bey, the owner of one of the two big hotels in Yusufeli, proposed that the town should be declared a disaster area so that residents can legally benefit from tax amnesties, debt cancellations, and direct cash transfers. Several shop owners who were particularly worried about losing their long-standing business relations insisted that they should be paid a lump sum that covers their possible future losses, as well as the costs for the transportation of their goods to the new settlement area. Later, the establishment of a technical school, gendarmerie barracks, or a satellite university campus in the new settlement area, and the construction of a highway that connects the town to Erzurum and other larger cities in the east would also emerge as key demands.

Especially after 2013, together with several contractors from the town, shop owners took the initial steps to unite behind these proposals. Led by Abdullah Bey, a wealthy lawyer and an active member in the local AKP branch, they eventually established the Association for the Protection of Yusufeli's Shop Owners (Yusufeli Esnafını Koruma Derneği) to coordinate the efforts to communicate the grievances of the town's "middle classes" to the DSI officials and high-ranking figures from the AKP. In the meantime, the mayor actively worked to extend the scope of these activities, especially during his second term after 2014. Lawyers, contractors, imams, and local representatives of almost every political party participated in a trip organized by his office in the summer of 2017 to Kurdish-majority Halfeti and Samsat, two towns already inundated by dam projects, and to Hasankeyf in southeastern Turkey, which, with the completion of the Ilısu Dam project, was about to be resettled. The municipality's website framed this trip as an "investigative excursion" whose aim was to draw lessons from the experiences of local residents displaced and resettled in other parts of the country. Participants were deeply shocked by flaws in the design of the new settlements areas they visited, which they described as "ghost towns" where business relations had suffered tremendously, and disturbed by the inadequacy of compensation payments, upon return they brought together a list of demands from the state. This list included several new proposals for revitalizing touristic and agricultural activities, increasing compensation payments, and planning new means of revenue-generation in the new settlement area, and also recommended the establishment of committees to collect information on the wishes and suggestions of local residents covering a wide variety of issues, including education, agriculture, village life, leisure, etc. Following up on the report, the mayor set up several local committees in collaboration with the local governor's office (*kaymakamlık*) to address these issues. Their findings would once again be compiled in a report to be sent to regional governors and AKP politicians.

Few ideas voiced by those residents who took part in local committees and by the members of the shop owners' association found official recognition. In addition to rejecting those requests either for some form of direct cash transfer or an increase in the amount of compensation payments, these politicians and bureaucrats were also criticized for failing to address the residents' questions in a number of critical matters. Eyüp Bey and the department heads of his office admitted that even they did not exactly know when the dam project would be completed and how expropriation rates be calculated. They also revealed that they were left in the dark about when the construction in the new settlement area was going to begin, which was a central issue for the majority. Residents anxiously waited to see some activity to commence in the Yansıtıcılar area as dam construction continued in full speed after 2014. In our conversations, the mayor attributed the dearth of reliable information to the lack of coordination between various local, regional, and national bureaucracies. He also expressed his belief that the shortcomings in the decision-making process could be overcome only under a more centralized political system where all power is relegated into the hands of a leader. His wish would come true soon when Turkey transitioned into a presidential system following a referendum in 2017, which received the support of more than seventy percent of the electorate in Yusufeli.[3]

In the meantime, another major concern in the town was *hak sahipliği* or who is entitled to relocate, and therefore be assigned a house, in the new resettlement area. According to the eligibility criteria set in the Resettlement Law (*İskân Kanunu*), residents of villages and towns subject to resettlement are required to pay to the related state department or agency, in Yusufeli's case to the DSI, an amount equal to eighteen times of the existing average wage, which was around 144,000 Liras in 2014.[4] The same law also specifies that only those families who will have lost their means of economic subsistence are allowed to relocate. Receiving social benefits from the state, in one way or another, in that respect, was interpreted by most residents to hinder their chances. Yusufeli's residents contacted regional MPs and cabinet ministers, through the mayor's office and the local AKP officials, to demand changes in the law as only a few residents in the entire town could meet these requirements. A temporary article added to the law in 2014 allowed the cabinet of ministers to make decisions related to resettlement issues. In June 2015, it signed a decision that invalidated the clause concerning the requirement about the loss of subsistence and reduced the fee for obtaining the right to resettle to one thousand liras, specifically for people who would be affected by the Yusufeli Dam project.[5] This was however not enough. The same decision reads that only those residents who had resided in the town center three years in a row after January 2011 would be eligible for owning a house. People who established businesses in the center after this date or are permanently or temporarily registered in villages would therefore lose their rights. Once again, following negotiations with state officials and ministers, Erdoğan signed a presidential decree in August 2019 that annulled the requirement to live in the town center.[6] A title deed or a workplace certificate that proves the resident's connection to the town would now suffice. Yet, there was still a caveat. Being married or "satisfying the condition for being a family" (*aile vasfına tabi olmak*), as the decree put it, remained the essential criteria for earning the right to relocate to the new settlement area. Hundreds of single young men and women were therefore either excluded or forced to apply only with their own parents to be able to start a new life in the Yansıtıcılar area.[7]

Hegemony and Its Limits

Bulldozer capitalism thus seems to have helped to produce and maintain support for the AKP rule in this part of Turkey in two ways: on the one hand, local politicians and prominent figures with a familiar face kept alive the possibility that residents' grievances related to compensation, displacement, and resettlement would be addressed and resolved. In a limited number of cases, on the other hand, the ministers or the president directly intervened to enact sovereign exceptions to existing rules and regulations to partially integrate their desires and anticipations into the interests of the ruling historical bloc. Forging long-term attachments and dependencies with citizens in both cases, the party-state successfully used incitements to negotiate construction and destruction as the cornerstone of its hegemonic project, without fully accepting, in the end, almost any of the major demands.

Several scholars interested in Turkey's construction- and extraction-based accumulation regime and its effects both on urban and rural communities have recently turned to Gramsci's *Prison Notebooks*. For example, one theoretical contribution (Akça, Bekmen, and Özden 2014) made the point that neoliberal hegemony in Turkey is constituted around strategies of accumulation led by a political project that represents the state as a national-popular entity to mask this institution's complicity in class-based domination. Writing more specifically on the hydropower projects in the Artvin region, Bengi Akbulut and her colleagues (2018) stressed that economic development and modernization become the core around which this national-popular outlook is maintained and reproduced. By animating development in the case of mega-dam projects as collective interest, the state surpasses different societal groups' particular demands and makes itself appear as a neutral or impartial actor to produce consent. While it reaches a more or less similar conclusion, Sevinç Doğan's (2016) ethnography of the local party structures and their mobilization in the Kağıthane neighborhood in Istanbul, by contrast, proposes to shift the focus from an abstract state or party apparatus to concrete figures. Mayors and some of the prominent party members of the local party branch, many of whom are intimately involved in the construction business in Kağıthane, mediate the articulation of different social classes' expectations and desires into the interests of the AKP. It is this capacity to represent the interests of the AKP as the general interest of the people, according to Doğan, that makes its project of hegemony successful.

Doğan's findings seem to point to a theoretical path already explored by Peter D. Thomas (2010) in his impressive reevaluation of Gramsci's work: the state in its integral form had become a network of social relations for the production of consent. Hegemony, understood as a new consensual political practice distinct from the use of force or coercion, in that sense, entails the capillary or permanent direction of an entire social fabric (2010: 143–44). Yusufeli's residents, I argued in this chapter, became "applauders" of their own destruction and displacement as a result of such a capillary action that helped to restructure residents' spatial and temporal sensibilities and experiences of ruination. Various actors from the party-state that embody and facilitate the connections between local and national scales build attachments to political power by creating the expectation that the terms and conditions of this settlement's submergence can be successfully negotiated.

Does all this entail a new kind of statecraft? Some of the recent ethnographies of real estate, infrastructure, and finance have argued that the political and economic relations woven around construction serve to sustain forms of state structures, which are in many ways radically different than those witnessed in previous periods. For example, Fabio Mattioli (2020) argued that the management of illiquidity through the delay of payments and suspension of deals in the context of massive urban projects helped to buttress the authoritarian Gruevski regime's hold on everyday anxieties and contradictions in Macedonia. Llerena Guiu Searle (2016) showed that the spectacular urban transformation in India made possible by intricate economic, political, and semiotic work done by local real estate companies, national politicians, international banks and investment firms helped to establish new connections between the public and the private to push the urban poor out of the land. Finally, Christine Folch's (2019) ethnography of the Itaipu Dam located on the transboundary Parana River skillfully studied the links between the processes of industrialization and electrification, and the ideas about and claims over national sovereignty in Paraguay. Especially in the course of the left-wing Lugo government's reign in power, people mobilized for the purpose of recovering the country's "hydroelectric sovereignty" by making various connections between nature and nation, as a result of which various actors expanded their economic and political powers in unexpected ways. My conceptualization of bulldozer capitalism as a system of power and profit predicated on the creation and annihilation of physical environment strongly resonates with several aspects of these accounts of statecraft. My material here on the governing of desires and attachments by the party-state through the processes of construction and destruction, and the role that conservative commons continue to play in the production of the consensual political practice, however, pushes me to retain the emphasis on hegemony.

This consensual political practice is not however without its own contradictions. The urgency of certain demands, as well as the conflicts within and between the representatives of the party-state, revealed, on some occasions, the fragility of the AKP's hegemonic project in Yusufeli as well.

The delay in the start of the new settlement's construction at a time when the building of the Yusufeli Dam project was going on was a constant source of complaint in the town. To be able to contain residents' criticisms shortly before the municipal elections in 2014, DSI sent two bulldozers, reportedly borrowed from a local contractor, to the Yansıtıcılar area to presumably begin the preliminary work for flattening the terrain. This, however, caused even more anger and frustration among residents as they saw it as a cheap tactic for convincing them into casting their votes for the AKP. During his election campaign, Eyüp Bey openly declared that negotiations for higher compensation and a better new town could continue only if the AKP would raise its vote share to sixty percent. Even though this thinly veiled warning sufficed to get him reelected, residents' fears and concerns continued to accumulate after the election to gradually culminate in quiet discontent.

Two protest meetings organized in Yusufeli in early 2015 should be considered in this context of the party-state's diminished capacity to manage the desires and fears of residents at this point. Coordinated by a new group called Yusufeli's Resurrection Platform (*Yusufeli Diriliş Platformu*), which included members and representatives from virtually all political parties active in the town including the AKP, they became the final occasions in which uncertainties around displacement, resettlement, and expropriation galvanized the activist energies. Yet, there were important differences between the two events. In the first gathering in January, thousands of people filled the streets to protest mainly against the amount of compensation announced by the DSI to be paid for the village land about to be expropriated; roughly 42 to 58 liras (€12 to €17) per square meter. After tossing two wooden coffins that represented the town center and the villages into the river, three thousand people continued their march towards the local DSI office. They chanted slogans and carried banners criticizing its bureaucrats for trying to take away their land for very little money, as well as failing to take shop owners' grievances into consideration. Some of my interlocutors who closely follow the developments around the dam project, as well as a number of contributors to a closed group on Facebook, took the view that this protest was little more than an attempt by the AKP to allow the residents to vent their anger in a controlled manner. By deflecting the responsibility from the party-state and its local and national representatives to DSI officials, they observed, the AKP could once again restore the conditions for making promises for a better future. The January protest also helped the mayor to reconsolidate his power both in the town and within the broader party networks. Eyüp Bey did not only quietly encourage the Platform members to organize the meeting but also very briefly marched with them on the main street. This act of political triangulation, it seems, enabled him to perform the role of a benevolent politician who on the one hand listens to his electorate's needs and grievances, while on the other hand serves as a mediator for the planners or national politicians by using their anger as a bargaining chip. Yet, as the remaining handful of anti-dam activists angrily pointed out, its end effect was simply to add to the normalization of the Yusufeli Dam project by making the displacement and destruction appear unavoidable.

The second protest organized by the Platform partially transgressed the limits imposed by such political calculations since it directly targeted an issue causing significant divisions within the local AKP branch: the location of the new settlement area. In summer 2014, Hilmi Bey, a local businessman and party member, began to lobby against the plans to move the town center to the Yansıtıcılar area despite the earlier legislation passed by the parliament. Arguing that this place would not be suitable for establishing a new town since its rough terrain and lack of water[8] would pose severe technical and economic difficulties, he publicly announced the village Çevreli, where he owns large parcels of land, as the best possible choice. To prove his point, Hilmi Bey personally introduced me to every single village head (*muhtar*), shop owner, and party member who shared his objections to rebuilding the town in the Yansıcıtılar area. After one interview, he even took me on his jeep to this hilltop to let me see with my own eyes the barren field filled with large rocks and asked me to report my findings to the "authorities," despite my attempts to explain to him that my research had completely different aims.

Whereas Eyüp Bey, his political allies, and the majority of residents continued to support the original decision, Hilmi Bey's Çevreli proposal was reportedly backed by several members of the party executive board in the town, bureaucrats from the DSI, and more importantly national politicians such as Kadir Topbaş. In one of his visits, Topbaş even reportedly scolded Eyüp Bey for insisting on the Yansıtıcılar area. Increasingly worried that Topbaş's involvement could tilt the balance of power in Hilmi Bey's direction, representatives of other political parties, together with some shop owners and village heads, who also took part in the first protest, once again mobilized the residents to take to the streets in March 2015. Even though other issues concerning expropriation payments were also voiced in this protest, the essential demand was that construction work immediately begin in the Yansıtıcılar area. Eyüp Bey abstained from taking part in this demonstration as he apparently did not want to be seen as openly challenging Kadir Topbaş and other prominent AKP figures. This controversy around the location of the new town was soon resolved after Erdoğan personally intervened and ordered the DSI officials to begin the construction in the Yansıtıcılar area. In 2016, the blueprints for the new settlement area were circulated in the town, a DSI office in charge of resettlement opened and shortly after bulldozers were sent to the hilltop, this time to genuinely begin erecting a settlement. In addition to a few changes made later in the design of the settlement area, such as the distance between residential areas and shops, this became one of the few tangible gains that came out of the negotiations with the party-state.

This chapter examined the relations of time and space to continue to explore what led the local community in Yusufeli to come to accept the plans for their displacement and dispossession. It discussed "looking forward" as the key temporal orientation that underlies the residents' accommodation to their valley's ruination. I looked in particular at the party-state's efforts to construct and colonize "looking forward," and how it underlies the consensual political practice that contributed to the rise of the AKP. I now turn to other extractive projects in valley and the different political responses that they elicited in and around Yusufeli.

Notes

1. This refers to the changes that took place during the Republican era, but the settlement's turbulent history under the Ottoman rule requires a brief overview. A sanjak (*sancak*) or an administrative division under the Ottoman Empire has existed in Kiskim, the old name for Yusufeli, since the beginning of the early eighteenth century. After the Russo-Turkish War of 1828–29, Kiskim was downgraded into a district (*nahiye*) and was subsequently integrated into different *vilayets*, including Erzurum and

Çıldır, throughout the second half of the nineteenth century. Especially, following the Russo-Turkish War of 1877–78, the difficulties concerning communication with and transportation to the area prompted the relocation of the district center from one settlement to another. The name Kiskim was officially changed to Yusufeli in 1914. For the only book-length historical study on Yusufeli, see Demirel (2010).

2. There seems to be methodological problems with this study as the majority of villagers are excluded from the survey. For the booklet, see *Yusufeli Yedinci Yerini Ararken* (Kamuoyu Araştrması 2012).
3. On 16 April 2017, a constitutional referendum was held to ask the electorate to approve or reject eighteen proposed amendments to the constitution, including the abolishment of the office of the prime minister and the introduction of an executive presidency. Won by the AKP and MHP, the original supporters of the amendments, with a very slight margin (fifty-one percent), under circumstances recognized by international observers as unfair, the referendum brought an end to the existing parliamentary system of governing. Recep Tayyip Erdoğan was elected as the first president in elections held on 24 June 2018, receiving the support of seventy-three percent of the voters in Yusufeli. Almost exactly the same number of people from the town had voted for the transition to a presidential system the previous year.
4. Between 2012 and 2018, the yearly average exchange rate between Euro and Turkish lira increased from 2.32 to 5.70 and continued to climb up to 7.92 by mid-2018. These currency fluctuations and exchange rate shocks compel me to provide the exchange value to Euro for every amount stated in lira for the year given. 144,000 liras was approximately equal to 50,000 euros in 2014.
5. For the full text of the decision (no: 2015/7877), see www.resmigazete.gov.tr/eskiler/2015/07/20150705-5.htm.
6. For the full text of the decree (no: 1480), see www.resmigazete.gov.tr/eskiler/2019/08/20190824-1.pdf.
7. For the politics of family during the AKP era, see Ayhan Kaya (2015) and Kocamaner (2018).
8. I will come back to this issue in Chapter 3.

— Chapter 3 —

Entangled Dispossessions

Losing All at Once?

In April 2013 in the city center of Artvin, I entered a small shop selling sports equipment to meet its owner İsmail, a high-spirited man originally from one of the villages of Yusufeli but who moved here in his teens. İsmail began the conversation by complaining about the expropriation of his family home in the village of Sebzeciler. This village was at the center of a dispute between its residents and the Doğuş Company at the time as it was about to be partly bulldozed to open space for the construction site of the Artvin Dam project.[1] But the conversation quickly moved first to the Yusufeli Dam project and then to Cengiz Holding's plans for extracting gold and copper in Cerrattepe. "What a fate!" İsmail quietly exclaimed, as if he were speaking to himself. "I'm about to lose first my village, then my town and now my city. At the age of thirty-seven, is this the sort of life that I'm meant to live?"

İsmail was not the only person I met whose multiple homes in the valley were being jeopardized at the same time by various resource extraction projects. Especially after it became clear by the mid-2010s that HEPP projects were planned in thirty-eight out of sixty-three villages in Yusufeli, hundreds of residents suddenly faced the prospect of witnessing their streams trapped inside steel pipes, in addition to losing parts of their settlements to the dam reservoir. Prospective mining and cadastral surveying took place more or less concomitantly and in some cases in the same settlements to trigger a wide range of political reactions. The majority of villagers from Yusufeli that I spoke to between 2013 and 2017 considered gold and copper mining as the only viable economic activity that they could take part in a future where the submergence of most parts of the valley and the displacement of their residents would have been concluded. But for the same reason, for them the prospect of losing their streams, already weakened by climatic changes, to energy companies had to be strongly resisted. Their participation in and expectations from various forms of bargaining and negotiation with the party-state and its local representatives that I discussed in Chapter 2, on the other hand, were no doubt also instigated by the effects of cadastral surveys. The forced transfer of meadows and agricultural land to the Treasury and the General Directorate of Forestry (Orman Genel Müdürlüğü) was completed for the most part without any visible opposition. Villagers knew, however, that they would not, as one interlocutor put it, "have anything much left to put on the table," once the expropriation would begin in the villages because of the Yusufeli Dam project.

In this chapter, I take the discrepancies between responses to capitalist extraction in this resource frontier as my starting point to highlight how bulldozer capitalism's reconstitution around a national-local nexus continues to generate both resistance against and accommodation to dispossession. I argue that political agency in environmental and resource-based disputes is formed relationally, emerging out of people's various considerations of what will be lost and what can be gained as different processes of destruction and expropriation become spatially and temporally linked. Entangled dispossessions, as an analytical lens that I employ here, intends to account for these calculations about the overlapping effects of extractive technologies, and in what ways they serve to amplify or hinder the capacity to get organized and articulate different societal forces against energy projects. I will show in Chapter 4 that the cost-benefit analysis of residents in the context of the Yusufeli Dam project ultimately gave rise to entrepreneurial ambitions, making them pursue material gains amid ruination. My intention in this chapter is to discuss how the contestations and negotiations around this project become conditioned by other processes of extraction and expropriation in the valley, and in turn shape their politics of dispossession.

Most residents living in and around Yusufeli immediately objected to the potential physical damage that the construction of small hydropower facilities would inflict on their settlements. The building of water canals made of cement and the use of dynamite for opening tunnels, they argued, would devastate the natural life and lead to soil erosion. Others said that they intended to continue growing fruits and vegetables on higher altitudes even after the lower parts of their settlements became submerged. If not enough water was left in their streams, however, there could be no gardening. It would not be possible, they added, to use the water for irrigation after it passes through the turbines.

But even in those places which are outside of the reach of the Yusufeli Dam project's effects or where little agricultural activity exists, residents still vehemently opposed the plans for the construction of HEPPs. For example, Tamer Bey, a leading figure from the campaign in Yaylalar, explained to me that pensioners and tourists came to his village to leave behind the hassle and noise of big cities. If the construction begins, it would not be possible to rest in his village. Located near the Kaçkar Mountains at the border of the Barhal National Park, Yaylalar is an idyllic mountain village whose population of five goes up to a hundred during the summer season. Originally a teacher who began to run a bed and breakfast some years ago, Tamer Bey was particularly worried that tourist activities in the village would be badly affected. For that reason, in addition to bringing a court case against the company behind the project in Yaylalar, he also contacted his village's co-locality associations from Istanbul and Bursa to make the campaign more visible. One weekend in summer 2016, hundreds of people from these two cities came back to visit their old village and to stage a protest in front of the stream where HEPP projects were planned to be implemented.

Asking what exactly motivates people to oppose the construction of small dam projects more enthusiastically and consistently than any other kind of energy infrastructure and how one should conceptualize these struggles given their diverse social, spatial, and political composition seems particularly interesting in the context of the middle part of the Çoruh Valley, given that its local struggles gained momentum at a time when the region's resources and commons were being expropriated and enclosed at a massive scale. Several scholars writing on the energy rush of the late 2000s have offered contrasting answers to what makes the opposition against HEPP projects unique, looking at the political, social, and sensory forces behind the unprecedented wave of protests that shook Turkey's countryside. Ceren Eren and Atakan Büke's (2016) class analysis of Fındıklı, for example, made the claim that the changing structures of agrarian capitalism and its unequal effects on the eastern Black Sea region explain why an exceptionally powerful anti-HEPP campaign came into existence in this coastal town. They argued that because tea, which is widely grown in small gardens in Fındıklı, has not been open to neoliberal market relations as intensely as tobacco and hazelnut—the region's two other major agricultural commodities—small tea farmers could afford to lead a precarious yet continuing existence. Together with their capacity to diversify their household income, the relative economic security afforded by tea farming enabled them to build a solid movement not vulnerable to the economic incentives offered by energy companies. Murat Arsel and his colleagues' (2015) study of the successful anti-coal struggle in the town of Gerze, on the other hand, has important implications for understanding the anti-HEPP struggles and for giving a name to the environmental politics of the 2000s: "environmentalism of the malcontent." Coined in reaction to the influential concept of "environmentalism of the poor" by Martinez-Alier (2002) and his tendency to attribute political agency in environmental disputes in the global South to economic impoverishment, this term captured the essence of the alliance formed between the residents of the villages where the coal plant was announced to be built and the left-leaning pensioners residing in the nearby city of Sinop. The scars of the state violence inflicted on generations of socialist activists, as well as the more recent anger directed towards the injuries of neoliberal developmentalism, according to this account, occasioned the convergence of societal forces to resist against this environmentally toxic project. Sinan Erensü's (2016) doctoral dissertation on the politics of energy infrastructures in Turkey was in part premised on the crucial but often neglected observation that after decades of outwards migration, urbanization, and decline in small-scale agricultural production, water can no longer be seen as a means of subsistence for the majority of the region's communities. According to him, rivers and streams should instead be considered part of the commons through which residents from across the urban-rural continuum breathe, enjoy, and express themselves; that is, replenish their labor power in summer vacations or heal the wounds of decades-long wage labor in Western cities. This suggests, Erensü claims, that it is social reproduction rather than economic production that plays a central role in giving direction to infrastructure-based disputes and confrontations, particularly in the context of the HEPP projects. Finally, Özge Yaka's (2017) feminist-phenomenological study put the emphasis on the corporeal connections with water and the nonhuman environment in explaining what instigates the wide-spread opposition against hydropower infrastructures. She argued that women's bodily, affective, and sensory experiences of nature and more specifically of rivers, in part conditioned by the gendered division of labor imposed by patriarchal structures, put them on the forefront of clashes and confrontations with the state and energy companies for defending their environment.

I intend to contribute to this discussion by scrutinizing how the links and tensions between the local and the national, and the town and the village render dispossession negotiable or insufferable. The political responses to entangled dispossessions in and around Yusufeli, I claim, take shape against the background of two important developments: on the one hand, the planning and implementation all at once of countless extractive projects in the villages make the interests, fears, and expectations of a plethora of actors to both converge with and diverge from each other, at times opening cracks within the party-state's governing apparatus. On the other hand, porous living arrangements that emerged between the village and the town in the past fifteen years or so after village residents began to relocate to the Yusufeli center, as well as the visible decline in agricultural production, transformed rural livelihoods in drastic ways.

Other Struggles, Other Calculations

By the second half of the 2000s, the local anti-dam/anti-displacement campaign organized around the Yusufeli Cultural Association began to turn its attention from the Yusufeli Dam to countless small-scale hydroelectric power plants (HEPP) announced to be built in the region. Zeliha began to visit the town more often around this time and developed a closer relation with Muhsin Abi and Hikmet Abi, who often referred to her in our conversations as "my teacher." A few years later, Ahsen would also join them from Berlin and eventually moved into a flat in the town center to devote herself to full-time activism for several months a year. Together with Hikmet Abi, these two women regularly traveled to Yusufeli's villages and those of the neighboring city of Erzurum to inform the residents about the projects planned in their settlements and to establish local contacts for leading anti-HEPP campaigns. By the early 2010s, lawyer Ragıp Bey began to represent those villagers who decided to take energy companies to court, often on the grounds that the projects were given a green light by the DSI without an EIA report; and Muhsin Abi kept regular contact with village heads, especially from Yusufeli, to receive the news about the latest developments. For a brief period of time in the early 2010s, Zeliha used the name "Çoruh Defense League" (Çoruh Koruma Birliği) to refer to this group and its efforts to save the region's streams. She also contributed to the planning of a large march held on Istiklal Street in 2015, coordinated by the members of various Artvin colocality associations based in Istanbul.

The activists' shift in focus was no doubt prompted by the withdrawal of the last international consortium in 2007, after which the mobilization against the Yusufeli Dam slowed down to be eventually rendered obsolete following the project's nationalization and the AKP's electoral victories. At the same time, the discontent growing against this type of hydropower technology in all corners of the country created a snowball effect from which Zeliha and her friends also tremendously benefited. Reportedly, the DSI's original vision by the end of the 1990s was to supplement the ongoing process of completing the damming of Turkey's largest rivers, including the Çoruh River, with the construction of a limited number of micro facilities on smaller rivers and rivulets (Erensü 2016). With the further liberalization of energy markets by the AKP in 2003 under conditions extremely favorable to capital however,[2] both small and large private investors, many of whom had no previous experience in the energy business, flocked especially into the eastern Black Sea region to embark on building several HEPPs on virtually every single flowing water source.[3] Locals' concerns about the transfer of rivers' use rights to private companies, as well as the decrease in the amount and flow rate of water, led to protests in almost every village, valley, and town in particular in this part of the country.[4] Also in Yusufeli, while the number of anti-dam/anti-displacement campaigners dramatically dwindled within a few years, anti-HEPP sentiments continued to escalate and led to the mobilization of residents in several villages where these projects were planned. Together with people from the town center and their co-locals living in Western cities, these villagers took legal actions and staged demonstrations to prevent or stop the construction of these facilities.

If there is one place that unmistakably became the driving force behind anti-HEPP activism in Yusufeli, it is Tekkale. As early as November 2011 when I visited this village for the first time, there was already a visible campaign going on against the planned project. Banners hung on village houses conveyed the message that residents would do anything in their capacity to prevent losing their stream. My host in Tekkale, Seyfi, told me that while the new village head had taken a bribe from the energy company in exchange of his support for the project, the majority of villagers were decisively against the construction of these small hydropower facilities. This had caused serious tensions over the years, culminating in occasional physical fights between the village head's supporters and other residents. Villagers took the energy company to court, arguing that an environmental impact assessment report should be prepared. While the judge ruled in their favor, this ruling was still not enough to decisively stop the company's plans, and demonstrations continued to be sporadically organized in Tekkale in the following years. In one protest action in the early 2010s, the villagers reportedly opened a banner that read "We will cast our vote but not give our water" (*oyumuzu veririz, suyumuzu vermeyiz*). What appeared to me at first as a cryptic slogan, was in fact something that lies at the heart of the intricate social and political relations that helped to maintain this struggle for several years.

Located a few kilometers to the west of Yusufeli, Tekkale is one of the closest villages to the town center. Possibly also for that reason, several people that I got to know over the years, such as a wealthy hotel owner, one prominent local AKP member and Muhsin Abi, have family connections to this place. From the beginning, these figures, including the board member from the local AKP branch, openly supported the efforts to stop the project, making possible the creation of a successful campaign whose support base cut across party affiliations. While I found this somewhat bewildering, given that it was the party and its leader who enabled the hydroenergy rush of the 2000s in the first place and repeatedly criticized or even vilified its opponents in public over years, pro-AKP figures from the village reasoned that there was little contradiction between supporting the party and defending their village against energy companies. Also encouraged by a few AKP politicians who shyly but publicly admitted that the construction of several small hydropower facilities on the same stream might lead to problems,[5] these residents did not hesitate to voice their criticisms against these projects. The slogan "We will cast our vote but not give our water" was, in that sense, a message directed to the AKP: they were willing to maintain their strong allegiance to the party despite their objection to the effects of its energy policies in their village.

Especially after 2013, first the cabinet of ministers and then the president took several urgent expropriation decisions to allocate village land to energy companies in different parts of Turkey where small dam projects were planned. The names of three villages from Yusufeli also appeared in a decision made by Erdoğan in 2015. As lawyers specialized in environmental disputes made clear, these decisions would necessarily result in the seizure of village land since they are, by law, exempt from the right of appeal. They can therefore be considered to operate outside of the purview of regular judicial norms, serving as legal exceptions used by the sovereign power, in this case the president, to remove the final barriers in front of capital's operations of extraction. Significantly, the land in Tekkale never became subject to an urgent expropriation order even though the HEPP project planned there is one of the oldest ones in the entire town.

There is enough reason to speculate that the mayor and other local AKP politicians might have intervened to save Tekkale. Shortly after he won the municipal elections for the second time in 2014, Eyüp Bey, in fact, asked Ragıp Bey, the former head of the Yusufeli Cultural Association, to represent all villages potentially affected by the HEPP projects in their legal fights against energy companies. He told me that the official view of his municipal office was that these projects are detrimental to nature, as well as the social and economic life in the villages. Yet, Tekkale was in a unique situation. Its stream, he continued, would provide the main water source for Yansıtıcılar, Yusufeli's future settlement area, and was therefore an absolute priority for legally challenging the energy company's plans. Eyüp Bey never confirmed in our interviews whether or not he personally contacted politicians from Ankara to ask for the favor of sparing this village's stream from damming. Also, even though the construction in Tekkale never commenced, the project has not officially been cancelled either. Despite all this, it seems that some form of sovereign exception deployed here for once might have produced the effect of blocking private energy capital's intrusion into streams because the interests attached to two separate extraction projects fundamentally clashed with one other. To make things more complicated however, mining companies too were waiting to profit from the same area between Tekkale and Yansıtıcılar, and I would learn about their long and convoluted history in the region in an unusual way.

Shortly before travelling to Yusufeli from Berlin in the summer of 2015, I received a phone call from Zeliha who warned me that a mining company was collecting information about me in the town. Apparently, my cautionary remarks about the plans for gold and copper mining in earlier interviews and conversations genuinely alarmed the company's local employees. Zeliha sounded worried enough to explicitly ask me to promise her that I would not travel to villages on my own. She even alluded to past events in which critics of dams and mining companies died under supposedly suspicious circumstances.

A few weeks later, accompanied by a friend from the town who wanted to make sure that I would be safe, I met the man trying to find out about me at the restaurant of a hotel in Yusufeli. Erkin was an athletic-looking middle-aged man who had earned a living as a tourist guide and a rafting instructor before he switched to working for various mining companies as a local contact. Far from posing a threat to me however, he quickly turned into an interlocutor who provided valuable information about various mining projects in the area. Erkin explained that in 2005 a Canadian company called Mediterranean Resources and its Turkish subsidiary, Akdeniz Madencilik, bought two properties, Taç and Çorak, situated on the northwest side of the Çoruh Valley, from Teck Cominco (later Teck Resources), another Canadian company, which was at the time also behind the project in Cerrattepe. Initial exploratory work conducted by Cominco had indicated that the area was suitable for the extraction of gold and copper. A report from 2008, prepared by a financial services company on behalf of this new corporation, which I found much later on the internet, confirms that the estimates of possible resources of gold mineralization amount to 2.3 million gold-equivalent ounces in these two deposits.[6] The same report also announces that feasibility studies in two new sites, one inside the Çevreli village and the other one near Tekkale, intersecting with parts of the new settlement area, began in 2007. I learned from Erkin that large gold deposits were eventually discovered in both places and the company had extended its exploration work to other sites in the neighboring Barhal Valley in recent years.

Erkin then revealed that in March 2015, a few months before our conversation took place, Mediterranean Resources sold all these properties to a Turkish mining company called Güriş. According to him, this was a result of pressure faced by international companies to withdraw from this sector in Turkey. As in small and large dam sectors, AKP was trying to ensure that mining business remained open predominantly to domestic companies on whose behalf it was ready to intervene. Even though the second Canadian company did not have difficulty obtaining exploration and operation licenses from the state, the permission to begin drilling for each new site was granted after a very lengthy review process. Also, it was asked to compile costly reports for operating the deposits in Barhal since they were partly located inside a national park. Güriş Company, by contrast, was allowed by the government to immediately begin its operations on all these sites, according to Erkin. It was also going to receive compensation from the state for the few parcels of land that it now owned, which were going to be inundated by the Yusufeli Dam reservoir.

The potential material effects of large-scale mining around Yusufeli are no less intimidating than those of small hydropower facilities proposed to be built on the region's streams. Even though neither Mediterranean Resources nor Güriş has specified the kind of mining technology that they intend to deploy in the area, local opponents of mining such as Ragıp Bey and Muhsin Abi were certain that the toxic substance of cyanide would be used to dissolve the gold from the ore. However, unlike small HEPP projects, virtually anyone else that I spoke to from the village and the town center strongly supported the project as they considered it to be the only long-term employment opportunity. Whereas the construction of small hydropower facilities is usually completed in less than a year after which there remains no need to employ anyone, except for one or two security guards, the prospect of continuously working for the same company or even in the same mining site for several years seemed promising. Erkin stressed in our conversation that Güriş had already built excellent relations with the local communities of the villages that would be affected by mining projects. Eyüp Bey ignored my concerns about the danger of environmental toxicity that mining gold with the use of cyanide could pose by simply stating that he did not see any harm coming from mining companies. He was not worried about the effects of land expropriation and physical destruction in the villages and the new settlement area either. Getting visibly annoyed, Ahmet Bey, his right-hand man in the municipality, even point blank asked me in another interview if I worked for a German NGO or a foundation, implying that my questions indicated I might be working as a spy for the German government.[7]

Dislocations of the Rural

In the meantime, cadastral surveys conducted in Yusufeli's villages mostly after 2010 caused widespread distress among its residents. As part of a plan that goes as far back as the early 1930s but nonetheless could not be implemented for several decades, the Turkish state envisioned registering boundaries of land ownership, which remained unclear in rural areas for generations, and recovering all state-owned land used by individuals. Outsourced to private companies across Turkey in the 2000s, cadastral surveying in the region entailed in principle comparing aerial images of the area taken in the 1970s with the data collected on the field to determine if residents had cleared parts of forests or grown crops on public land in the past forty years. As my interlocutors bitterly complained, however, in practice the employees from the surveying company automatically recorded those parcels of land situated above water canals of villages as the property of either the General Directorate of Forestry or the Treasury. This included commons such as meadows that the villagers collectively use for grazing, but also uncultivated agricultural land was often registered as public land. Especially the latter decision put the residents in a difficult position as it reduced the amount of property that they anticipated to turn into cash when the expropriation of their villages would begin because of the Yusufeli Dam project.

A close interlocutor until 2015 and the brother of my imam friend whom I mentioned in Chapter 1, Acar Bey, a civil servant then working at the Yusufeli municipality, told me that roughly two thirds of all land in his village Irmakyanı was gradually taken away from the residents after 2005. While the villagers could previously benefit from the meadows and threshing fields, as well as other forms of unregistered land, on the basis of use rights passed from one generation to another (*zilliyet hakkı*), the cadastral surveys implemented by the state ignored this practice to transfer the property of these parcels either to the Forestry or the Treasury. Taking me to Irmakyanı in his car one late afternoon in April 2013, Acar Bey grew visibly enraged as he continued to point to me one site after another that sits above Irmakyanı's water canals and thereby was registered as state land. He appeared equally helpless however, as he reminisced about moments and events from his childhood connected to these places while lamenting in the same breath that their losses would translate into less money once the state begins the expropriation of their village because of the Yusufeli Dam project.

Irmakyanı was one of the first villages in Yusufeli where this form of dispossession has taken place. Because a construction site was already planned to be built in the middle of this village by the mid-2000s when the Doğuş Company got involved in the Yusufeli Dam project as part of the second international consortium, the planners might have opted for resolving some of the property issues before buying village land. Cadastral surveying that took place in Yusufeli's other villages after 2010 also led to the loss of large parcels of land and elicited reactions similar to Acar Bey's. In Tekkale, for example, villagers complained in 2014 that the employees from the surveying company, some of whom were their own relatives, transferred almost seventy percent of all village land to the Forestry and the Treasury. Given the absence of title deeds or any other legal documents through which they could claim ownership, their chances, they told me, of winning court cases against these two departments was very low. In the meantime, Erdinç Bey, a friendly employee from the private surveying company that I briefly spoke to in his tiny office in the Yusufeli town center in summer 2014 appeared completely unaware of the villagers' grievances. Rather than discussing the implications of cadastral surveys on people's lives however, he instead wanted to show me the state-of-the art computer programs that he used to register the dispossessed land. The interviews that I conducted in the offices of the Yusufeli District Forestry Management (Yusufeli Orman İşletmesi) over the years shed more light on the ways that different forms of dispossession become enmeshed in one another, and what sort of state logics accompany their implementation. All three employees from the District that I spoke to clarified that, contrary to what some villagers claimed, cadastral surveys were not at all related to the Yusufeli Dam project, with the partial exception of Irmakyanı. As a matter of fact, by the mid-2010s the Çoruh region was one of the last places left in the country where the recording of land ownership was still not completed. Murat Bey from the District expressed some sympathy for the villagers losing their lands and meadows to the state through cadastral surveys at a time when they were also grappling with the effects of the Yusufeli Dam project and struggling against the energy companies' plans for building small hydropower facilities. However, despite his own reservations, he readily admitted that he would not dare to show leniency to villagers using forest land. In addition to being liable to fairly strict forestry laws, he also had to take into consideration that these parcels of land could be used for energy investments. Because some of the water pipes and almost all electricity transmission lines planned in conjunction with dam projects pass through forest land, his department was also in charge of preparing environmental impact assessment reports which influence the decision on whether or not to transfer the use rights of forest land to energy companies for forty-five years. As another employee, Emir Bey, added, no one from the District would "try to act like Don Quixote" to block an energy investment. Responding to my questions about the villagers' chances of legally challenging the outcomes of cadastral surveying, Emir Bey solemnly ruled out the possibility by repeating the phrase "one cannot sue the state" (*devlete dava açılmaz*). Whereas some villagers had apparently found some success in retrieving the land transferred to the Treasury, not even a single person could win a court case against the District Forestry Management in Yusufeli. As an explanation, he once again referred me to a number of articles from the forestry law that stipulates that even if no trees grow on a particular parcel, it may still qualify as forestry land and therefore may be registered by his office. "Besides," he continued, "there will be no such thing as a village within the next two or three decades anyways." People would soon spend time in rural areas only for leisure activities. Perhaps, he was not wrong after all.

Like in many other parts of the region and the country, Yusufeli's villages felt almost abandoned during the winter months, bar the presence of a handful of elderly people. In summers however, village populations swelled by ten or even twenty times as migrants residing in cities returned to spend their vacations there. While people continued to move to cities in western Turkey, the majority chose to relocate to the Yusufeli center in the course of my research. These families led a dual existence between the town and their villages as they kept their houses and land. One major reason for this decision was the closure of vital institutions such as health clinics in villages. Also, following an educational reform, which increased the duration of primary education to eight years, village schools began to shut down at the end of the 1990s and the schools in the town center became these families' only option for sending their children to school. New construction projects in the valley, as well as the AKP's routine use of İŞKUR[8] for temporally employing people, helped to absorb some of this new labor force throughout the 2000s. As several of my interlocutors noted, this growing detachment from the economic life in villages intensified the new town center residents' expectations to turn their properties in villages into cash. Therefore, in addition to the "urban-rural continuum" that Erensü (2016) draws attention to in order to explain the protests against small dam projects in the northeastern parts of Turkey, one can also speak of a "town-village continuum," which, at least in some cases, hindered the possibility to maintain the wider alliance built against the construction of the Yusufeli Dam project.

Yet, things were not much different for those who stayed either. Only a very small minority of villagers must have been truly dependent on subsistent agriculture in contemporary times in the entire middle part of the Çoruh Valley. This is primarily related to the geographic characteristics of the valley and how they shaped agricultural practices. Because there are no flat plains in the valley, people have often created small terraces, not exceeding a few *dönüm*, on the hills by carrying the soil up from the riverbed. But thanks to the heavy sedimentation of the Çoruh River and the Mediterranean microclimate of the area, it became possible for them to use this extremely fertile soil to cultivate almost any kind of fruit and vegetable that grows in Turkey[9] with very high yields every year (Kurtiç 2019). Beekeeping, animal husbandry, and the cultivation of grains were more common in villages either bordering forestry land or situated on higher altitudes. Both the very small size of the land and the productiveness of the soil required relatively little labor power and agricultural inputs such as fertilizers. Cultivation of cereal grains such as rice and rye were more labor-intensive and dependent on state subsidies. A big portion of agricultural products were sold in the markets in the neighboring city of Erzurum. While most villagers had other sources of income, such as pensions, remittance money sent by relatives from Western Europe or wages earned by means of working as civil servants, workers, or artisans, agriculture still provided an important supplementary income that most families used for covering the school expenses of their children.

This form of production that interlocutors described to me alternatively as semi-subsistent or part-time farming, depending on the type of the crop and the location, began to unravel by the early 2000s. By the time I arrived in Yusufeli, most villagers' reliance on market relations for their own consumption had already become more pronounced. Friends living in the town often complained that these people, including their own relatives, purchased not only tea, sugar, and oil but also eggs, meat, vegetables, and fruits from the supermarkets at the center. The future submergence of villages, as well as the decades-long uncertainties about its timing, were mentioned as some of the immediate reasons behind the decline of agricultural activities. Villagers hesitated to continue cultivating their terraces since they repeatedly heard from the state authorities that the expropriation of their villages would begin in the near future. But also, the effects of the Çoruh River's heavy damming became visibly detrimental for agricultural production. Ongoing construction of roads and tunnels, which often involved dynamite to explode rocks, produced clouds of dust and sand that traveled across the valley to cover fruit orchards. Agricultural reforms implemented in the wake of the November 2000 and February 2001 economic crises as a result of the standby agreements that Turkey signed with the International Monetary Fund (IMF) too no doubt affected this region.[10] After existing agricultural subsidies were phased out, restrictions on agricultural products were imposed, state-owned agricultural institutions were privatized, and the agricultural bureaucracy was permeated by a broader technocratic market rationality (Aydın 2009; Keyder and Yenal 2013; Doner 2015); small-scale or family farming has been decimated across the country. Especially the decline in animal husbandry around Yusufeli was attributed by villagers to increasing costs of inputs such as livestock feed and the elimination of subsidies. With the partial exception of beekeeping, which continued to find support from development agencies, rural production has been practically reduced in less than two decades to almost a niche activity.

An organic boutique shop opened on the main street of Yusufeli in 2016 offered a wide range of village products including nuts, honey, olive oil, cheese, eggs, milk, and *pestil*[11] to visiting tourists. My friend Atakan, in the meantime, imagined renting the few *dönüm* of land that his family owned in their village to urban tourists in summers for "hobby gardening." Many villagers continued to keep one or two animals. These were mostly bulls that they rigorously looked after by feeding them the highest-quality fodder that they could afford, as well as Viagra, for entering them into the series of bullfighting festivals organized in spring and summer every year. And while Erzurum was no longer a market for fruits and vegetables grown in Yusufeli's villages, several people built greenhouses on their land starting from the second half of the 2010s. As we will see in Chapter 4, however, this was not primarily related to agriculture either.

Notes

1. After it was forced to withdraw from the Yusufeli Dam project, together with the second international consortium, in 2017, Doğuş Company won the tender for the construction of the Artvin Dam.
2. Several changes made in the rules and regulations concerning the energy market since the early 2000s have rendered the hydroelectricity sector especially very profitable from the point of view of capital. One legal milestone that preceded AKP's rise to power was the introduction of the Electricity Market Law (No: 4628) in 2001, which made private capital's intrusion into energy production easier by enabling the issuing of licenses to companies. This law was slightly amended in 2013 (No: 6446) to eliminate the speculative license market that emerged after 2001, replacing small investors with big companies. The Water Usage Agreement was accepted in 2003, on the other hand, to provide the legal basis for the construction of small dam projects by private companies. With this agreement, parts of rivers and rivulets were leased to private companies for at least forty-nine years. Finally, amendments made in the law concerning administrative procedures in 2014 made it more difficult for judges from regional courts to decide in favor of stopping the execution of energy plans. For more information on the legal basis of the energy rush of the 2000s and 2010s, see Erensü, Evren, and Aksu (2016: 13).
3. As of 2018, the total number of small and large dams in the Artvin region is fifty-six. Whereas thirty-two of these projects have already been completed, the rest is either in the process of construction or planned to be built within the next few years.
4. Two important groups that sought to bring these local campaigns and struggles together were *Derelerin Kardeşliği Platformu* (Sisterhood of Rivers Platform) and *Karadeniz İsyandadır Platformu* (Black Sea is Rising Platform). For more information, see Aksu (2016).
5. For example, in 2016 the Minister of Forestry and Hydraulic Works, Veysel Eroğlu, had to publicly admit that he shared the concerns about micro HEPP projects, which produce less than 10 MW of electricity.

6. This report is no longer available in the company's website. After the 2008 economic crisis, its shares dropped to three cents to apparently trigger the removal of some members of the executive board and the restructuring of its mining operations. For more information, see Allan (2011).
7. The conspiracy theory that the local struggles against mining activities in Turkey are directed and funded by German civil society organizations such as the Heinrich Böll Foundation gained widespread acceptance among nationalists after a local campaign against gold mining started in Bergama, İzmir in 1992 and received national attention within a few years. Especially one book written by historian Necip Hablemitoğlu strongly made the point that the discovery of gold deposits in Bergama posed a threat to the German state as Turkey was importing large quantities of gold from this country. For the Bergama struggle, which is one of the earliest environmental struggles in Turkey, see Arsel (2016b).
8. İŞKUR or *Türkiye İş Kurumu* (Turkish Employment Agency) is a public institution established in 1946 for the purpose of decreasing unemployment. As with many other state institutions, under the AKP governments it was turned into a partisan tool strategically utilized to temporally absorb parts of unemployed masses, in certain locations, and especially before elections.
9. One important exception however is tea. Unlike most parts of the eastern Black Sea region, humidity levels stay relatively low because of the Mediterranean microclimate of the valley and this makes the cultivation of tea impossible.
10. For an early and detailed article on the effects of IMF-imposed structural reforms on agriculture in Turkey, see İslamoğlu (2002).
11. *Pestil* or *köme* is a dried layer of fruit pulp consumed as snack. It is usually made of grapes in Yusufeli.

— Chapter 4 —

Economies of Construction and Destruction

Speculating on Ruination

Yusufeli began to bear the marks of a truly booming town already in summer 2013, shortly after the foundation stone of the dam project was laid. Hundreds of new buildings, already built or in the process of being constructed, were expunging the last traces of a rural feeling in the town center. Shortly after, new footbridges, walking tracks, and small parks were completed, as dirt roads were covered with asphalt and riverbanks were paved with concrete blocks. Each of my stays in the following years helped me to discover another layer of this remarkable material transformation and its manifestations in the lives of Yusufeli's residents. Notices about new building projects hanging in the doorways of pharmacies, hairdressers, and Internet cafés promised the comforts of modern apartment life. New supermarkets, clothing shops, and restaurants popping up on a regular basis competed with one another to cultivate a desire for conspicuous consumption among the more affluent inhabitants. I remember discreetly observing an elderly man trying pizza, apparently for the first time in his life, in a café recently opened to cater to the new tastes of the town. Occasionally, I would also indulge in a watery cappuccino in the same place, instead of drinking black tea in one of the traditional coffeehouses scattered across the town. Especially in the early evenings, when company buses returned to bring back their passengers—workers and engineers from dam, road, and tunnel projects—Yusufeli would feel more like a busy working-class district of Istanbul than a mid-size town in the eastern Black Sea region.

Figure 4.1. Construction boom in the Yusufeli town center, 2013. © Erdem Evren.

For outsiders, it was not easy to comprehend that what was gradually appearing in front of their eyes, almost on a daily basis, was destined to disappear without a trace within a couple of years. A well-known anthropologist colleague from Germany who heard about my research later revealed, almost apologetically, that he had spent a few days of his summer vacation here, entirely oblivious to the town's imminent fate. The Canadian mining company, whose long-term plans for extracting gold and copper from Yusufeli's two villages, which I discussed in Chapter 3, addressed the ongoing real estate boom on its website with a short text published alongside photos of half-finished buildings arising across the town. Unsure about how to make sense of this development, its unknown author openly asked if this was an indication that the dam construction was about to be delayed yet again.[1] Chain-smoking on sleepless nights early in my research, I too occasionally felt possessed by the temporal dissonances engendered by the economies of construction and destruction permeating the town all at once. The neon sign of a new supermarket, "Centenary" (Yüzüncü Yıl), illuminating our balcony from across the street with the perplexing statement, "Yusufeli will reach its centenary with us" made me wonder: if so, then will the town not be submerged after all? When I posed this question to Centenary's cashier, she remained unconvinced, but was unable to explain how anticipations of the future could find a place and flourish amid conditions irreversibly moving in the direction of obliteration.

This chapter follows residents' lives after they moved from trying to prevent the loss and destruction of both things and ways of living, to investing in their dissolution as a source of valuation. I examine how their desires and expectations to capitalize on the disappearing land, property, and social relations fueled projects and visions based on rendering the valley's devastation commensurable with long-term investments. From buying or creating property with the aim of receiving higher compensation payments, to developing ambitious plans for selling artifacts bound to vanish with the submergence, residents developed different strategies to profit from the ruination of their surroundings. I examine some of the political forces and circumstances that enabled them to actively tap into the valley's devastation in order to potentially make economic gains in the future. I also discuss what kinds of new social and economic divisions and hierarchies were engendered and reproduced in the town.

Ironically, if one essential condition for preventing the disappearance of the town and its villages was to block the tempos and flows of dam capital, as I argued in Chapter 1, at this point, making calculations on the built environment's demise and, specifically, how it spatially and temporally unfolded became required for taking part in extracting value from it. Residents are constantly interpellated to a future-oriented existence marked in particular by speculative ways of thinking and acting, which made me turn my ethnographic attention once again to instances of "looking forward" and similar temporal sensibilities invested in the "yet-to-come." The point I make here is that, like negotiation, speculation became after 2012 one of the prevailing tendencies in Yusufeli to navigate through the material effects of the valley's heavy damming, and the uncertainties and opportunities they elicited in the lives of the inhabitants.

While some of the ideas and plans for profiting from the effects of the valley's devastation and future submergence had already appeared in the earlier phases of the Çoruh Energy Plan's implementation, it was only following the AKP's victory in the municipal elections in 2009 that they became substantially widespread. Unlike many other places where physical or legal violence was deployed to implement infrastructural projects, the conservative-nationalist identity and predominantly pro-Erdoğan loyalties of the community in Yusufeli obliged the use of tools and mechanisms that sought to produce legitimacy and compliance for managing the politics of displacement. Circumventing or intervening in existing administrative frameworks at the local level to turn the slow-motion destruction of the town and the valley into a conduit for residents to take up plans for accumulation became the party-state's essential strategy. In this chapter, I focus in particular on the interventions by the mayor and his office, who, in coordination with national party networks and administrative bodies, bent or modified existing legal frameworks and regulations, especially those concerning property and compensation. These temporary and contextual paralegal arrangements (Chatterjee 2004) enabled the residents to participate in the real estate boom by boosting the prospects for profiting from the future compensation economy.

But then a broader entrepreneurial spirit, partly exceeding the effects of these arrangements, took root and spread at around the same time, prompting residents to pursue economic security and class mobility through speculative projects of one kind or another: turning material and immaterial ruins into commodities, performing dam labor as a form of investment in future livelihoods, or entering into intricate circuits of financialization. My ethnography documents indebtedness and precarity as the most common predicaments in Yusufeli, which are driving ordinary people to adopt this entrepreneurial ethos, shaped under the economies of construction and destruction. While these conditions are no doubt shared by billions of people around the world who endure different faces of early twenty-first century capitalism, this bulldozer variant appears in my account as particularly successful in generating and maintaining long-term political loyalties amid crisis. The transmutation of ordinary people's expectations about the future into ways of thinking and acting rooted in speculation and financialization engendered social and economic dependencies that facilitated governing without resort to force.

Land, Property, and Speculative Accumulation

After nearly two decades, the ban imposed by state agencies on new buildings in the town center looked entirely obsolete by summer 2013. Even if a few wealthy residents and at least one engineer working for the municipality had managed to circumvent this ban, which had been imposed by the DSI to minimize the amount of compensation payments, the majority had been dissuaded for all those years from repairing old buildings and infrastructures, let alone motivated to build new ones, with the official justification that the dam construction would begin within the next few years. My interviews with the mayor and two different contractors confirmed rumors that construction permits for about a thousand buildings had been issued in the previous two years. Four-, five-, and even six-story blocks with central heating and elevators were rising on every corner. Rents for most of these modern flats were roughly triple the amount typically paid for older ones, and land prices soared at similar rates, especially in the town center. This inflationary tendency was echoed in everyday life, as small carpenter workshops expanded into spacious furniture stores, and ordinary shops turned into fancy household appliance stores and hardware stores, all under the watchful eyes of fellow residents. The real estate agent training classes offered by the local community center (*halk eğitim merkezi*) were also packed during this time, just like those classes on how to use and maintain drilling machines and bulldozers. Young and middle-aged men and a few women flocked into these classrooms (some of which were directly financed by construction companies), to become licensed truck drivers, machine operators, and real estate brokers, hoping to benefit from the new opportunities materializing in the town.

Urgent demand for both short- and long-term accommodation, especially among young families, engineers, and migrant construction workers, was visible in the town during my stays. However, my conversations with contractors and new property owners revealed that a desire to make a profit on the future compensation economy was the primary motive behind these purchases. Locals were investing in real estate based on the calculation that the rents they would collect, together with the payments they would receive after the expropriation of their property, would exceed their initial outlays. Many of these new property owners and builders were among wealthier inhabitants who had enough capital in the first place to pursue these long-term investments. More interestingly, however, a significant number of villagers whose land and houses had already been expropriated for the Deriner and Artvin Dam projects also bought flats in Yusufeli around this time. By using the sum they had received from the expropriation of their assets to invest in real estate in Yusufeli, they pursued a kind of speculative accumulation that Jeremy Campbell (2014: 240) characterizes in his ethnography on Brazilian Amazonia's recent frontier settlers as the attempt to "predict and perform future shapes of social and economic infrastructures."

Yet, it was not only the town itself that was witnessing a construction boom. Residents of three neighboring villages showed me the greenhouses and barns they had recently built and told me of similar projects that their relatives or acquaintances from other villages had embarked on. The same logic was also in effect in these villages. The construction of greenhouses and barns, the villagers explained, mainly served the purpose of creating assets that could be liquidated in the form of compensation payments in the future; they were not intended to establish permanent infrastructure for agricultural production and livestock rearing. Just like the new property owners in the town center, these villagers were also speculating on the coming submergence of their settlement by participating in its economies of construction and destruction.

Katherine Verdery's (2003) seminal work on the effects of decollectivization in a rural Transylvanian community remains a useful reference point here for making sense of the devaluation and revaluation of land by Yusufeli's residents. Similar to what I argued in Chapter 3, Verdery discovered that land's value in primarily economic terms became diminished during the early postsocialist era as a result of the increasing costs of inputs and the overall changes affecting rural life, turning farmers into what she calls "owners without efficacy." Social values related to status and mastery, however, compelled the same people to continue working their land for a time, until many of them began renting their parcels to tenants or "supertenants." In the villages of Yusufeli, by contrast, the impossibility of making land valuable through agriculture under the particular conditions of dam construction left little room for a revaluation in moral terms. While the hectares in Eastern Europe eventually became the cornerstone of a rentier economy—even as they vanished both in material and moral senses—the *dönüm*s of Northeastern Turkey were reduced to objects of entrepreneurial desires predicated on speculating on destruction.

I had the chance to learn more about these villagers' anticipations when a young resident from Arpacık, a village located a few kilometers south of the town center, accepted my offer to visit him on one hot summer day in 2015. Mesut was growing fruits and vegetables in his garden and in the greenhouses that he set up on his ten-*dönüm* land that he had inherited from his father. However, he did not seem too concerned when we found out that perhaps one-fifth of his crop had been burnt under the August sun because illness had prevented him from coming to water the field in the previous couple of days. Explaining that about eighty percent of all irrigated land in his village would be lost after the completion of the Yusufeli Dam, he responded to my initial questions about his purpose in continuing agriculture with the quip, "the dam is approaching, let's see the values!" Mesut was particularly optimistic about the amount he was expecting to receive for his greenhouses, claiming that the state would pay fifty thousand liras (16,000 euros) for each *dönüm* of greenhouse, whereas it cost him about half that to build them in the first place.

His main complaint at the time, however, concerned a piece of legislation that had recently been passed to postpone the delivery of expropriation payments for five years. While in previous cases in the valley cash payments were transferred to the property owners' bank accounts within a few months after the expropriation started, this period was dramatically extended in the case of the Yusufeli Dam project in order to prevent a mass outward migration that would turn Yusufeli into almost a ghost town before the completion of the project. Therefore, even though a private company subcontracted by the DSI had already begun to document the properties subject to expropriation and make estimations on their values in some of the villages around the town center, including Arpacık, the residents would have to wait for years before receiving their money. More importantly, once the assets of each family that were subject to expropriation were

Figure 4.2. Greenhouse in Arpacık, Yusufeli, 2015. © Erdem Evren.

determined by the company through close-range photogrammetry and satellite imaging, no payment would be made for enhancements to those properties within the five-year waiting period. Mesut likened his predicament to "some kind of slavery" as he was prevented from building new houses or greenhouses on his land, since these enhancements would not be compensated by the state. In fact, another villager that I spoke to on the same day added that, while he started to build a greenhouse in order to receive more money from the state, he was "caught by the satellite" (*uyduya yakalandım*) before closing its top and therefore jeopardized his prospects for receiving full compensation.

As I would find out, things were no different in the town center. At least since early 2013, DSI officials who oversaw the dam project expressed their objection to the construction boom, which was drastically increasing the amount of compensation payments to be paid to the residents. Through his connections to cabinet ministers and important politicians from the party, the mayor of Yusufeli, Eyüp Bey, had successfully delayed bans on new construction several times. A ban finally went into effect in summer 2015 when satellite images of the properties designated for expropriation in the town center were taken and began circulating. Officials from the DSI made an announcement around the same time that owners would be compensated by the state only for those properties that appeared in the images. Unfortunately for some contractors and property makers, several of their houses were only half-finished when these images were taken. According to Eyüp Bey and other residents, about three hundred building projects in the town center were documented through satellite imaging before their completion. Complaining about being caught by the satellite, these residents expressed worries about their chances of receiving full compensation and immediately contacted the mayor's office. After a period of uncertainty, the mayor once again convinced the AKP politicians from Ankara to intervene and ensure that these buildings, as well as barns and greenhouses, became subject to full compensation. In his own words, he "saved those people who legally received construction licenses but then went against the law."

The mayor and his office also actively intervened to inflate land values and to redraw the borders of the municipal area. This had to do with the fact that parcels of land within the municipal borders were expected to make more money compared to those in villages. In one case that a former civil servant described as a form of legal skullduggery (*hülle*), the municipality finalized the sale of a parcel of land worth 350 liras (around 55 euros) per square meter for 1000 liras (around 200 euros) in order to artificially set a higher market value. The mayor was hoping that this would strengthen a property owner's hand in future negotiations with the expropriation office of the DSI when they made the case that parcels of land within the municipal borders were more expensive than what these officials were willing to pay. Similarly, by changing the legal status of lands adjacent to the town center—that is, by designating them as municipal land—the mayor sought to increase the value of those properties and therefore expand the scope of land and real estate speculation in the town.[2] Unlike the non-Indigenous settlers in Jeremy Campbell's (2014) account, who created property in the absence of the state in the remote area of Castelo de Sonhos as a means of anticipating the future order of governance, the speculative strategies for accumulation by residents living in and around Yusufeli emerged from below; but it operated and flourished strictly under the patronage of a political party that came to represent the state on the local level. What seemed to be at stake here was not so much foreseeing the future shape of rules and regulations as making temporary and contextual legal arrangements (Chatterjee 2004) to incite Yusufeli's locals to tap into the economy of things that were about to disappear. I will come to this point later.

Nonetheless, this entire schema of investment around property and real estate remained extremely risky. First, according to current regulations of expropriation and compensation, a significant sum—based on the depreciation rates calculated according to the building's age—was deducted from what was supposed to be paid to property owners. Second, because the property taxes due for each immovable asset were proportional to their value, homeowners faced the challenge of paying to the state a significant percentage of the rent they collected in this heated real estate market. Property taxes skyrocketed, especially after 2013, when the mayor's office began to inflate land prices in the town center. But more significantly, given the history of dam construction in the valley, no one, including the planners, could confidently estimate when the project would be completed and when the expropriation would officially start. On the contrary, in 2013 and 2014 the majority of investors admitted that there was no guarantee that a profit could be made in the end, and they openly acknowledged that they would lose money if the construction of the dam took more than six or seven years. At several points in my research, new property owners asked my advice as to whether they should wait until the beginning of expropriation or sell their flats to other residents immediately. The researcher with his contacts to politicians and company engineers, they assumed, would know how to time their investments.

I had the chance to understand better who came out as the winners and losers of speculative accumulation only toward the end of my research. In October 2017, after more than half of the dam construction was completed and the first houses began to appear in the new settlement area, some residents from the town began to contact the expropriation office to finalize the selling of their properties. In a handful of cases that I closely followed, those modern flats built after 2011 and sold by contractors for around 100,000 liras (44,000 euros) were purchased by the DSI for approximately 70,000 liras (17,000 euros). These property owners received an additional ten to twenty thousand liras, depending on the size of their land. Even though the construction of houses in the new settlement area was still going on at this stage, based on these figures, several of my interlocutors made the following prediction for an average Yusufeli resident who intended to resettle in new Yusufeli: about half of his or her compensation payment would automatically be deducted to cover the prepayment for a new flat in the new town, announced by the DSI to cost 117,000 liras. To be able to pay for the rest, he or she would be obliged to obtain a long-term mortgage from a bank. Even though these loans were expected to have a low interest rate of 0.5 percent, they would still end up with a net loss.

The risk of long-term indebtedness looked even grimmer for those residents who were unwilling or unable to take part in speculating on property and real estate. More than two-thirds of all shop owners were renting their shops and therefore were not eligible to receive compensation. Hundreds of residents faced the risk of being paid half of what would be paid to the owners of new flats, depending on the size and the age of their buildings. For example, Muhsin Abi, who owned five flats and one shop in one of the older buildings right at the center of the town, bitterly complained, after much calculation, that all his compensation money would barely be enough for to get him two flats in the new settlement area. Contractors or affluent residents who owned modern flats, on the other hand, were bound to emerge better off.

I reached a similar conclusion concerning the expropriation process in the villages. In summer 2017, the DSI officially began to purchase the land in Arpacık and two neighboring villages. Even though he did not want to share with me the exact amount estimated to be paid to him for all his properties, Mesut told me that he and his fellow villagers were offered around 60 liras (14.5 euros) per square meter of land. Given that he owned ten *dönüm*, one-third of which had greenhouses on it, and a house, I calculated that his total compensation package probably exceeded 700,000 liras (170,000 euros). While this looks like a lofty sum, it was nevertheless going to be shared with his four siblings, reducing his part to one-fifth of this figure. Probably also for that reason, he decided to decline this offer and instead took the DSI to court in an attempt to increase his compensation offer.

The mayor, prominent members of the local AKP branch, and predictably all lawyers in Yusufeli strongly encouraged the residents to resolve the matter in court, rather than accepting the offers made by the DSI's expropriation office. Some of the villagers whose properties were expropriated after the completion of the Deriner and Artvin Dams had had some success in raising the initial amounts they were offered by bringing the DSI to court. In Arpacık, however, virtually all villagers except Mesut quietly accepted their compensation packages, which was usually less than one-fifth of what he was offered. It was usually those villagers who owned large parcels of land or those without any debt who had the opportunity to take this route. In another village, called Dereiçi, one resident successfully doubled the initial offer of 50 liras (12 euros) per square meter by going to the court. Overall, the average two or three *dönüm* of land that most villagers living around Yusufeli possessed helped at most to cover existing debts, even if they had built greenhouses and barns to receive more compensation money. At the same time, the expropriation of their properties turned them into long-term debt-encumbered homeowners in the new settlement area.

Partha Chatterjee (2004) is therefore correct in observing that, in places where violence cannot be an option for displacing communities for purposes of extracting value from the environment, temporal and contextual arrangements made by the state, especially around property and land, inform a new kind of governmentality. My material, however, demonstrates that it is neither the actual rehabilitation of the dispossessed nor the mobilization of the subaltern in political society that underlies the modality of governing at stake here. Paralegal arrangements made by the mayor and his office rather helped to generate expectations, often of a fictional nature, that profit could be made out of the demise of the built environment, livelihoods, and ways of living. To a certain extent, comparable in form to sovereign decisions made to allow corporations to appropriate resources, these localized forms of legal exceptions encouraged the community to invest speculatively in the built environment's coming annihilation. As we have seen, however, with the exception of a few contractors, lawyers, and large landowners, these investors ended up with losses and suffered from further indebtedness.

Selling Ruins

Other forward-looking strategies invoked what Jane Guyer (2007) calls "fantasy futurism"; interestingly enough, they do so to the extent that they draw from the past and its remnants-in-the-making as their raw material. Material artifacts, ways of life, and emotions attached to the landscape and the built environment that were bound to disappear with the completion of the dam projects were intended to be packaged as commodities for future consumption. In that sense, the plans and visions that I came across over the course of my stays aimed at making a profit after the submergence by bridging timelines.

Some of the engagements with time that I witnessed earlier were markedly different from these profit-seeking strategies, even if they may have been embedded in another sort of fantasy futurism. In summer 2012, shortly before its submergence by the reservoir of the Deriner Dam, some of the activists from Artvin involved in the Green Artvin Association's campaign against plans for gold mining in the Cerattepe area buried a letter in the garden of a *kümbet* near the village of Sirya, which is also known as Zeytinlik.[3] Together with another *kümbet* and the Ferhatlı Bridge built by the Seljuk Empire, as well as the Berta Bridge, believed to have been completed in the late nineteenth century, this dome-shaped mausoleum left by the Saltukid dynasty sometime around the twelfth or thirteenth century was among the most important examples of architectural heritage in the entire valley. Activists were particularly furious about the planners' inexplicable decision to allow the two bridges to become inundated after completing their renovation in 2010. The letter they penned and hid in the garden began as follows:

> Dear *Türbe*,[4] dear future,
> I hereby list the names of the state departments and the people who allowed you to be flooded and drowned under the water. They are directly responsible for your drowning. I report them to you and to the future.[5]

The letter went on to narrate the process by which these structures ended up being inundated below the dam's reservoir, and it listed the names of twenty-seven people and their ministries and state departments, including Recep Tayyip Erdoğan and the then Minister of Culture, Ertuğrul Günay, as the culprits. With this action, activists told me, they intended to document an act of injustice and its perpetrators to which they bore witness. In the absence of a political will or an agent that could hold them responsible for the loss of cultural heritage in the contemporary moment, ruins here were animated to construct a politics of remembrance for the future. Past, present, and future were intended to be bridged in order to call into action the next generation—"the people who will discover these historical artifacts in the future," in the words of one activist—to hold these politicians accountable for the injury they committed against the valley, its history, and its people.

On my subsequent visits to Sirya after 2013, the village that I also described in the introduction to this book, the village head Eşref Bey insisted on showing me the artifacts that he spared and carefully amassed in one corner, away from the reservoir of the Deriner Dam: a wooden cradle; rusty utensils that were once used to produce olive oil and wine; an old television set; and a radio. He explained to me that he was planning to exhibit these objects in an ethnographic museum that he hoped to open in the new village settlement. "We are going to turn our new village into the Bodrum of the Çoruh Valley," he solemnly repeated in our conversation, referring to the famous holiday destination in the southwestern part of Turkey. We entered

Figure 4.3. Old winemaking artifacts in Sirya, Artvin, 2015. © Erdem Evren.

one of the few empty houses that had remained entirely intact, as its wealthy owner living in west Turkey had not seen the need to sell its parts. Looking out of the window that opened onto the large reservoir, Eşref Bey commented that it felt as if one was inside one of those waterfront houses (*yalı*) sitting along the Bosporus.

Shortly after the expropriation of the village land concluded, Eşref Bey had officially applied to Artvin's special provincial administration (*il özel idaresi*) to start the arduous process of reestablishing Sirya a few hundred meters above its current location. The historical mosque of the village had already been carefully dismantled and reconstructed in its new spot next to the Artvin highway. However, while he was allowed to borrow bulldozers from the provincial administration to flatten the land designated for the new settlement area, the last time I spoke to him in the fall of 2017 he was still waiting for the official permission and the financial aid required to begin the construction. It remained unclear even later in my research whether the parcels of land in the new place would be transferred free of charge to the village legal entity on behalf of former villagers, or if it would cost about half of the compensation money paid for their properties in the old village. To my slight surprise, Eşref Bey strongly preferred the second option, as he wanted the new Sirya to be populated by affluent residents rather than poor families, a few of whom were still living at the time in the half-flooded village inside small prefabricated huts provided to them by the governorate of Artvin (Artvin Valiliği). Despite all these legal uncertainties, however, he remained hopeful about establishing a seaside resort–like village with an ethnographic museum, which could bring in much-needed cash after all the village's agricultural land and locally famous olive groves were lost to the reservoir. Carefully keeping an eye on every small truck arriving in Sirya to remove the remaining junk, he continued to collect those artifacts that he saw fitting for his imaginary museum, investing in a future where lost objects and memories would one day acquire commodity value.

My friend Haydar, the older son of Muhsin Abi and the new owner of the local photo studio, similarly told me about his intentions to sell "anything and everything that will remind the residents of the old Yusufeli" after the submergence. Throughout the years that I visited the town, Haydar consistently struggled to keep his business running, which was hit not only by exchange rate shocks and inflationary spikes but also by the proliferation of digital cameras and the spread of online shopping in Yusufeli, just like the rest of the country. In addition to taking portrait and wedding photos, and trying to sell cameras, he gradually began to make a profit from his side business of printing landscape photos of the valley, which his father and he had taken over the years, on merchandise such as mugs, T-shirts, and clocks. "One makes more money out of the past," he announced to me several times, as I watched him diligently retouching images on his computer. After nearly four decades, the studio's negative archive had accumulated sixty thousand items. He planned to digitize and catalog this archive, mainly comprised of portrait photos of the town's residents, including dead ones, to be able to sell them via Facebook once the resettlement began. He asked me to check English-speaking photography forums to find out if there existed a scanner that could automatically digitize photo negatives.

Yet, it was the mayor of Yusufeli who clearly had the most ambitious projects in mind regarding packaging timelines. As part of his broader plan for turning the new Yusufeli into a hub for what he called a "tourism of longing" (*hasret turizmi*), Eyüp Bey specifically mentioned the idea of building in the new settlement area a "Minia Yusufeli": a miniaturized version of the town as it exists right before its disappearance, which would allow its visitors to walk among and touch its scaled-down buildings and streets. In one of our interviews, he wondered aloud how much the residents would be willing to pay to be part of this experience. In a later interview, he enthusiastically shared with me another idea of creating a three-dimensional tourism platform in which people could virtually move through the old Yusufeli. With the aid of video images, music, narration, and sound effects accessed with goggles and headphones, he was planning to offer the customers—old residents and tourists—the chance to retrieve in virtual reality the sensations of living in Yusufeli before its disappearance.

Eyüp Bey's fantasy futurism should in part be considered a desperate attempt to create revenue-generating schemes under economic conditions bound to become even more dire with the submergence of the town. He was deeply aware, perhaps more than anyone else in the town, that after nearly all agricultural land gets swallowed by the reservoir and the vibrant business environment becomes stultified as a result of the resettlement plan, which was expected to lead to a substantial decrease in the number of shops and restaurants, tourism of one kind or another would be the only option for making money. Both Eyüp Bey's office and some residents like my driver Orhan had officially asked DSI officials on a number of occasions for permission to organize commercial water sports activities such as kayaking and canoeing on the future reservoir. This request was promptly rejected on the grounds that the dam lake would not be safe for recreational activities. In the absence of any other viable economic plan, other than willfully succumbing to the expanding frontiers of mining in the region, creating commodities around longing and nostalgia was what Eyüp Bey could afford to offer to his fellow residents.

Anchored in the musealization, digitization, and virtualization of the ruins, then, the three distinct imaginaries of Eşref Bey, Haydar, and Eyüp Bey adhere to other residents' strategies for creating or buying property to place an unusual kind of entrepreneurialism at the center of the efforts to endure the disappearance of land, community, and the past. Predicated on speculating on the future exchange value of those objects, places, and relations destroyed or about to be irretrievably exhausted by capitalist development, this entrepreneurial subjectivity is key to understanding how ruination becomes articulated with accumulation amid social, environmental, and economic devastation.

The pickers and traders of the matsutake mushrooms, whose lives Anna Tsing (2015) tracks in her widely praised ethnography, do not have a choice, either, other than to survive the ruins. Long after the erosion of the welfare state and in the absence of regular wage labor, seeking collaborative survival in ruined environments became the only option for many of them. Tsing refers to this mode of production as "salvage capitalism," under which contamination of human and more-than-human lives, and the encounters between traders and pickers from different ethnic groups in places like Open Ticket Oregon, turn into a condition for producing value. In the middle part of the Çoruh Valley, by contrast, it was a profit-seeking subject taking risks based on a cost-benefit analysis by responding to the interventions and promises of the party-state that endeavored to endure the effects of bulldozer capitalism. With the exception of Haydar's fantasies with regard to integrating his business into the circuits of platform capitalism, the state remains vital here for creating value out of ruins and ruination, as it facilitates people's speculative and futuristic visions to potentially make economic gains. As we will see now, the state also remains the tutelary of local desires for class mobility and employment.

Investments of Labor

Evidently, not all residents had the means to speculate on land and property or develop ambitious plans for extracting value from the remnants-in-making in the future. Especially after the construction of the Artvin Dam began in 2010, countless new tunnel, road, and viaduct projects began to pop up in the middle part of the Çoruh Valley. Finding a job as a worker, driver, or security guard in one of these building sites became the most attractive employment opportunity for lower-class residents. To these young men, a construction job promised security that encompassed a low but regular wage, health insurance, and an entitlement to an old-age pension. But perhaps more importantly, work in construction was also seen as an investment insofar as it provided the opportunity to learn new skills transferable to other building sites in other parts of the valley, region, or the country. By acquiring the knowledge to operate or maintain a construction machine or a vehicle, these unskilled residents of Yusufeli, often sons of villagers and shop owners, could begin to envision a place for themselves in the real estate and infrastructure-based booming economy of Turkey.

This sense of improvement was widely echoed by the inhabitants that I spoke to early in my research. A man in his early thirties who took me to his old car in August 2012, responded to my questions about the effects of the Yusufeli Dam project on his life by professing the merits of dam building in the valley. "Dams are good for Yusufeli," he told me point-blank, since his fellow locals and he could learn a new profession that they could practice in other places. At the time, he was working at the Artvin Dam construction site, which was then expected to be completed within the following two or three years. With his new skills, he anticipated that he would have no trouble finding employment in one of the other construction projects along the valley.

Even some of those residents who remained critical of the town's submergence for electricity production would gradually be swayed by the promise of a construction job. Ali, the young owner of a local kebab shop that I befriended quite early in my research, was very critical of the Yusufeli Dam project, even though he had almost no contact with the local association behind the anti-displacement campaign. One day in the summer of 2012, I found him arguing in front of his small restaurant with a man who had recently found a job as a driver in the Artvin Dam project. Mehmet explained to Ali and me that he had already learned so many things in a couple of months at the building site, which would not have been possible to learn in school. Besides, even if the Artvin project were quickly completed, he would most certainly find a job in the Yusufeli project, this time as a security guard. He was also sure that Yusufeli would become a center of attraction (*çekim merkezi*) after new roads and tunnels connected the new settlement area to other parts of the region. Tourism would develop thanks to the reservoir, and the residents would have more job opportunities. "I see all this as a plot concocted by gold mining companies against Yusufeli," Ali firmly retorted. Repeating the conviction common among anti-displacement activists that the heavy damming of the Çoruh River was motivated by the companies' desires to extract gold and copper rather than by the need to produce electricity, he described the Yusufeli Dam as "a project aimed at depopulating the valley." Besides, there would be no job guarantees for anyone in the future. At best, Yusufeli's wealthiest would get richer while others became poorer. Five years later, when I bumped into him in the town center, Ali had long closed his shop, tried his luck as a waiter in a nearby city, and now was working as a construction worker at a tunnel project near Yusufeli. "I'll be working there for the next three or four years," he calmly announced to me. "Tunnel, viaduct, and dam projects will continue to go on, I can always find a new job." Even if construction work did not require many skills or teach new ones, perhaps with the partial exception of dam building, it still helped to keep alive the hopes of building a career in this sector for young men like Mehmet and Ali, many of whom were either previously unemployed or working in small family businesses.

For Yusufeli's residents, one essential condition for getting a job on these building sites was to ask for a reference from a prominent member of the local AKP branch. Well-known sympathizers of the opposition parties were in principle excluded from this labor market, while AKP voters considered a construction job more or less the expected remuneration for their support of and loyalty to the party and its leader. I remember feeling slightly surprised when Hatice Abla, a middle-aged woman who became one of my closest friends in the field, revealed to me that she half-jokingly threatened Kazım Bey, a wealthy shop-owner and a well-connected figure from the party, that they would not get her family's vote in the upcoming elections if they did not place Murat Abi, Hatice's husband, in a job in the Yusufeli project. Hatice Abla was trying to make ends meet with the tiny clothing store she was running from the lobby of the old building that she was living in with her husband and two kids. She had complained to me several times about her husband, who had begun to sit idly at home or kill time in coffeehouses after he closed his small tea shop. Murat Abi's reluctance to subscribe to the command and control of labor, however, was successfully circumvented a few weeks later when he was abruptly hired as a driver for one of the shuttle buses taking workers to the construction site every morning.

But again, even AKP's tight control over hiring practices to benefit those residents loyal to the party was not enough to satisfy everyone's desires, as three members of a youth association made clear to me in April 2014. Established in October 2013 with two hundred members, this association had the aim of convincing the companies and the local authorities to increase the number of workers from Yusufeli in the dam construction sites, as well as to find solutions to what they perceived as discrimination towards them. The association demanded rent support from the state, as the average rent of 700 liras (240 euros) per month was above what the majority of ordinary workers could afford to pay. Even though construction companies hired about five hundred residents for the Artvin Dam project and more than a thousand for the Yusufeli Dam project, these young men still expressed their discontent with the fact that the majority of workers were migrants coming mainly from eastern provinces. They accused in particular LİMAK, the biggest company that is part of the consortium behind the Yusufeli project, for bringing in large numbers of Kurdish workers instead of hiring exclusively from the Çoruh region. According to them, whereas these experienced workers were assigned to physically less-challenging tasks, such as working with construction lubricants or in the hydraulics section, the unskilled workers from Yusufeli were given more arduous jobs for less money. One young man from the association said that he was asked to work as a cement mixer driver for eighteen hours nonstop while others were normally required to drive for only eight hours. Also, it was the workers from Yusufeli who were often laid off in times of economic slowdown or crisis, since they were considered to be unskilled and easily replaceable. As a matter of fact, at the time of our interview two hundred workers from the Yusufeli Dam were staging a sit-in in front of the construction site to protest against their imminent dismissal.

All these complaints and grievances were narrated in extremely nationalist language that revealed the extent of the residents' anti-Kurdish racism, exacerbated by fears about the uncertain future and the challenging work conditions. "People living in southeast Turkey own large plots of fertile land," the head of the youth association claimed toward the end of our interview, with a sense of anger that he did not even try to conceal. "And the state sees what those people are doing against [our country]. The state is about to displace us. It is going to deprive us from our past and future even though it knows [that we are loyal citizens]. Is that fair? Is that fair that someone coming from Samsun [from the middle part of the Black Sea region] or

Diyarbakır [the largest Kurdish city] can work in dam construction, whereas people from our villages remain unemployed?"

In fall 2013, these strong anti-Kurdish sentiments had culminated in a spectacular act of violence when residents attempted to lynch Kurdish construction workers employed at the Yusufeli Dam site. Earlier hostilities between workers at other construction sites in the valley were usually quickly contained before turning into serious assaults. In this case, however, an ordinary fight between two workers that reportedly started after one of them tore an Atatürk poster into pieces in the workers' dormitories escalated into a lynching attempt. Hundreds of men from the town began to walk to the dormitories near the construction site, demanding that all the "culprits" be handed over to them. After the police forced them to return to the center, they attempted to raid the local hospital, where the injured Kurdish worker had been taken. It took the local governor (*kaymakam*) several hours to convince the residents to leave the premises and go back to their homes.

The young dam workers who spoke to me anticipated that the valley's protracted destruction by large infrastructure of different kinds would facilitate the building of new livelihoods at a time when older ones were rapidly disappearing. They hoped they could achieve some form of economic safety and class mobility by turning to party networks, but they ended up becoming new additions to the country's expanding reserve army of labor. Often lacking in any skills, unlike the Kurdish workers, from which Turkey's construction economies have historically drawn its labor power,[6] Yusufeli's young proletarians could only lead precarious existences at the bottom of formal and informal workplace hierarchies. A deeply rooted sense of ethnic supremacy combined with their inability to realize the promises that the valley's protracted destruction held in terms of employment opportunities fueled their racist violence against Kurdish workers. Among other things, it is these intricate articulations between class and ethnicity, or the subtexts of class hidden underneath the headlines of nation,[7] that explain why it was a lynching attempt, instead of working-class solidarity, that materialized in Yusufeli.

When Ruination Meets Financialization

Ultimately, though, bulldozer capitalism is composed of several linked moments by which capital is reinvested and recirculated. Many circuits of money have appeared over time along the Çoruh Valley to contribute to the redrawing of class maps in Yusufeli. Reinvestment of expropriation payments or profits made from the construction boom eventually helped to sustain a rent economy to the benefit of a handful of wealthy inhabitants. But then, a debt economy, in some ways peculiar to this part of the country, also began to take shape around the same time and to trouble the majority of residents. In this section, I briefly turn my attention to broader mechanisms that tie finance to the economies of construction and destruction.

Several people who became contractors, or *yapsatçı*, during the construction boom in the town were originally owners of shops specialized in selling household equipment and devices. At a time when the Central Bank of Turkey (CBT) continued to sink interest rates to historical lows, these shop owners applied for investment loans to be able to finance the construction of their buildings. As the term *yapsat* (build and sell) implies, several of these flats were quickly purchased by people who had either received enough cash after the expropriation of their properties in other parts of the valley or had obtained mortgage loans from Halk Bank and Ziraat Bank, two state banks also active in Yusufeli. Whereas shop owner-cum-contractors tremendously increased their wealth by selling washing machines, sofas, or television sets to these new flats, new homeowners quietly waited for the beginning of expropriation in the town center. In some cases, they also rented these flats to other families to collect rent money. In addition to real estate, the construction sites of dam projects also provided lucrative business opportunities, especially between 2012 and 2017. Some wealthy individuals bought construction vehicles; in particular bulldozers, around this time with the intention of leasing them to construction companies such as LİMAK for thirty thousand liras a month. Described to me as a particularly profitable investment, it apparently took less than a year and a half to cover the initial outlay. Other businessmen bought shuttle buses similarly for the purpose of renting them to dam companies, which were in dire need of a way to arrange the transportation of workers and engineers between the town, the villages, and the construction sites. In both cases, fuel costs, as well as the salaries of the drivers, were paid by the companies behind the Artvin and Yusufeli Dam projects. As a matter of fact, the most colorful rags to riches story told to me several times over the years involved a local junkman who bought the first shuttle buses for the Artvin project after he befriended an employee from the Doğuş Company and established close contacts with wealthy members of the local AKP branch. In just a few years, some of my interlocutors complained, he ended up becoming one of the richest individuals in the entire town.

On the other side, less well-to-do villagers and shop owners tried to enter this rent economy initially by applying for loans at the local branches of two other state institutions: Agricultural Credit Cooperative (Tarım Kredi Kooperatifi) and Shop Owners' Credit and Surety Cooperative (Esnaf Kredi ve Kefalet Kooperatifi).[8] Originally specialized in providing financial support to farmers and shop owners, both cooperatives became instrumental during the 2010s in buttressing these residents' visions of profiting from the real estate boom and the future compensation economy. Low-interest loans issued by the Agricultural Credit Cooperative, normally based on the condition of expanding agricultural activities in villages, were used by its inhabitants either to build barns and greenhouses or to buy houses in the town center. Several shop owners who did not own their properties and therefore were not eligible for receiving compensation payments, contacted their Cooperative about receiving loans to be able to purchase mortgaged houses and rent them to others. The rent money that they collected, however, was often enough to pay only for their daily expenses or for discharging existing debts. To be able to make their monthly mortgage payments, they once again applied for loans, this time to Halk Bank. Interestingly, to be able to secure loans from the bank, they used for collateral those properties that would be expropriated because of the Yusufeli Dam project, often an old family house or a few *dönüm* of land in the villages. Apparently, this was allowed only by the bank branches in Yusufeli. It became, however, such a common practice that, to the locals' surprise, the Halk Bank branch in Yusufeli received from its headquarters for several years in a row a "Blue Star Certificate," which indicated a "high operational quality," essentially meaning higher profits.

Neither the drastic increase in private household debt nor the constant struggle to replace old liabilities with new ones was unique to Yusufeli. The debt-fueled financialization of the economy, made possible by the AKP government's ability to keep interest rates both for personal and commercial loans at low levels, lay at the heart of the economic growth witnessed in the country from the beginning of the 2000s onwards. Ordinary people, many of whom were already indebted or precariously employed, could envisage investing in property thanks to the party-state's redirection of low-interest loans issued by banks and cooperatives to sustain the country's construction boom. The investments in ruination that I discuss in the case of

Yusufeli can therefore be seen as metonymic of a broader financialized subjectivity.

Some ethnographers have observed that the financial crisis of 2008 made it difficult for companies or individual contractors to complete construction projects in the peripheral countries because of the global lack of credit (Mattioli 2020). Turkey, however, was among a handful of developing countries that tremendously benefitted from the massive amounts of money created by northern central banks via quantitative easing to control some of the worst effects of the crisis. Despite the early attempts by the Central Bank of Turkey and the Banking Regulation and Supervision Agency to prevent rapid credit expansion and a sharp increase in household debt (Akçay and Güngen 2019), the capital that flowed from northern countries quickly boosted the Turkish economy, in particular the construction sector. Erdoğan, who announced already in 2009 that the global crisis would pass at a tangent to Turkey, intervened in critical moments to keep interest rates low enough to facilitate the borrowing schemes of non-financial companies. As a matter of fact, as the figures that political economists Ümit Akçay and Ali Rıza Güngen (2019) compiled from the TMB data indicate, both foreign capital inflows and economic growth recovered enough to reach pre-crisis levels in 2010 and 2011. The amount of domestic bank loans also tripled between 2008 and 2013, and remained stable for the next five years to continue sustaining infrastructural and real estate projects; meanwhile, Erdoğan sacked a total of five central bank governors who dared to contravene his open policy of invigorating the construction economy through low interest rates.[9] Starting from the second half of the 2010s, the promises that this debt-economy held began to unravel as one exchange rate shock after another compelled the TMB to increase interest rates in order to be able to continue attracting foreign capital. Erdoğan, however, insisted that the rates should stay either the same or go even lower so that the finance-induced construction boom could continue. Despite its figures being heavily manipulated by state agencies, according to most economists, the drastic rise in inflation, in turn, exacerbated the level of indebtedness and economic hardship.

Visiting Hamdi Bey, the head of the Shop Owners' Credit and Surety Cooperative, in his office in the course of my fieldwork, gave me the opportunity to witness some of the effects of this shift in the town. An avuncular figure who took a liking to me over the years, he enjoyed sharing with me his observations on the Turkish economy and the changes going on in Yusufeli. Already in 2013 and 2014, Hamdi Bey expressed his concerns about the increase in the non-recurring debt loaned by his Cooperative. In the summer of 2016, he revealed that out of five million liras given to shop owners, a total of 450,000 liras was never paid back. Whereas one third of borrowers, half of whom had already declared bankruptcy and gotten their properties seized, stopped communicating with Hamdi Bey and the Cooperative, the rest were barely surviving by applying for new bank loans to be able to cover their initial debt. "When I loan them money, I don't care if they open a new shop or buy a house or a car," he said. "Most of them, however, buy houses as an investment, to turn the crisis into an opportunity (*krizi fırsata çevirme*), so to speak. But if the house costs sixty thousand liras, two-thirds of it is paid with the loans that we issue. Because they are mortgaged to us, the Cooperative will receive its money back once the state begins to expropriate these houses and pay us compensation."

This chapter studied in three parts how people participated in economic projects and visions based on rent and speculation, which ultimately threw the majority of them further into debt and precarity. I showed how this happened under specific conditions where the slow-motion destruction of the Çoruh valley did not fuel opposition to the Yusufeli Dam project, but instead set the conditions for strategies of accumulation and commodification. Speculating on land and property, selling or preparing to sell the valley's ruins and ruins-in-making, and investing in construction labor and machinery were made possible in many cases by paralegal arrangements and interventions, as well as by broader financial mechanisms, orchestrated by the party-state and its local networks. Through their immersion in rent and debt economies as such, Yusufeli's residents, I argued, developed loyalties, dependencies, and affective attachments to the AKP regime. This dynamic relation remains one of the less explored aspects of how authoritarian formations like the party-state in Turkey continue to generate consent among those communities on which it wreaks social, economic, and ecological havoc.

A notion of entrepreneurship, if taken in an expansive sense, can therefore be said to encompass the majority of projects and visions of accumulation and commodification that emerged in and around Yusufeli in the course of the 2010s. From the shop owners and contractors to the workers and villagers, the majority of residents took or envisioned taking risks, financial or otherwise, in the pursuit of profit by exposing different aspects of the ongoing ruination to a cost-benefit analysis. My ethnography in this chapter documented in particular those spatial and temporal relations, as well as individual and collective desires and anticipations, produced by the calculations made to navigate the uncertainties surrounding the material devastation awaiting this community.

Is this economization? This term, as it is well known, refers to those actions, devices, and descriptions qualified and assembled as economic by social scientists and market actors (Çalışkan and Callon 2009). While earlier conceptualizations considered in particular the ways in which economic knowledges and materialities play a performative role in the making of markets, more recent contributions perceive the economization of the social through the naturalization of economic processes and the technocratization of their governance as that which distinguishes neoliberalism from other forms of capitalism (Madra and Adaman 2014). The economization of social relations and the politicization of problems that are deemed to be in need of economic solutions, it is argued, provide authoritarian regimes with the ideological tool of developmentalism, helping them to establish electoral legitimacy (Arsel, Adaman, and Saad-Filho 2021).

My material on Yusufeli suggests that the future-oriented work that went into making a profit out of the destruction and submergence of Yusufeli and its surroundings is closer to Laura Bear's (2020) understanding of speculation than economization. In Bear's brilliant anthropological reworking of the concept, speculation proceeds by "making value uncertain and then projecting unseen ethical orders using technologies of imagination that can help navigate this uncertainty" (Bear 2020: 3). Importantly, racial, gendered, or class-based differences that give shape to these imaginings of the social both permeate and are reproduced by speculation. While Bear and her colleagues are in particular interested in exploring the ethical orders and claims related to the histories of colonial and postcolonial extraction, the technologies of imagination that govern the social relations in Northeast Turkey are shaped by the ethnoreligious hierarchies and inequalities that inform the entire history of the country, as well as more recent polarizations connected with the rise of the AKP.

One question remains, however: what are the consequences of leading a life in Yusufeli in anticipation of its demise? To be able to answer this question, I now turn to those narratives and practices in which the residents employed the conceptual opposite of "looking forward," that is, "looking backward." Following their cue will allow us to grasp a different relation between accumulation and ruination, mediated by violence, whose effects continue to reverberate in the valley in the present.

Notes

1. Once again, this text was later removed from the company's new website.
2. In one case, the value of a large parcel of land near the Yansıtıcılar area, recently included within the municipal borders, was appraised by the mayor's office at 250 liras per square meter, but the DSI brought a court case against this decision.
3. Most villages in the region have two names: an old and often Armenian or Georgian one, and an official Turkish one. In this case, both names are used widely by the residents.
4. *Türbe* is another name for shrine or mausoleum.
5. For the letter, see Emen (2012). Even though my translation is based on the version that appeared in this newspaper article, I also interviewed two people involved in this action. The other *kümbet* was first covered with cement and then left to be inundated under the reservoir of the Deriner Dam.
6. While there seems to be no academic work written specifically on the ethnic dimensions of construction labor, leftist publications from the 2000s and 2010s are full of news stories about lynching attempts against Kurdish construction workers, especially those working in the northern and western parts of Turkey. At the same time, the forced migration that Kurdish families have been subject to throughout the 1990s is argued by some scholars to have resulted in the Kurdification of labor in Turkey in recent years. For this, see Yörük (2009). For its critique, see Benlisoy (2012).
7. Here, I am alluding to the title of the important work *Headlines of Nation, Subtexts of Class* by Kalb and Halmai (2011).
8. TESKOMB is the abbreviation for the national organization's name.
9. Interest rates began to increase starting from the second half of the 2010s, and the policy interest rate set by the Turkish Central Bank reached 19 percent in 2021.

— Chapter 5 —

TENSES OF VIOLENCE

Miserable Child Locked in the Basement

Especially in the beginning of my research, virtually all residents described Yusufeli as having been an exceptionally abundant, peaceful, and secure place before dam constructions began in the valley. I observed firsthand the extremely fertile fruit and vegetable gardens, formed from the alluvial soil carried by the Çoruh River, that give four or five yields a year; but in earlier times, meat was also cheap and widely available, usually served in a popular dish called *kavurma*. Shop owners, I was told, used to leave their goods unattended overnight without any fear of theft. The local prison had to be shut down by the authorities by the end of the 1980s, since it had never held more than a handful of inmates, most of whom were incarcerated for minor offences such as writing bad checks. Perhaps more remarkably, the political violence that erupted in the second half of the 1970s dividing neighborhoods and towns across Turkey along sectarian and ideological lines hardly touched Yusufeli. Left-wing activists living in other parts of the region attributed the town's exception in this sense to its provincialism or residents' predominantly rightwing worldview. One close friend from here who grew up in western Turkey but returned years later to work as a teacher went so far as to compare Yusufeli prior to the 1990s to the small American town depicted in Lars von Trier's celebrated film *Dogville*. His cynical remark did not mean to allude to the abuse and manipulation the female protagonist is subjected to in the second half of the film. It was, rather, intended to stress the town's remoteness and uneventfulness. Having repeatedly come across local narratives in which violence appeared only in reference to its absence, I must have asked myself around the same time who the miserable child locked in the basement could be.

The child I had in mind is a figure from Ursula Le Guin's (2000) famous short story "The Ones Who Walk Away from Omelas." The story's unnamed narrator depicts Omelas as an idyllic city where the happiness of its inhabitants depends on the perpetual misery of a single child kept under terrible conditions. Once citizens are old enough, they are taken to a broom-size closet in a basement to be shown the tormented child. While most of them eventually decide to stay in the city after some questioning, a few residents cannot come to terms with this awful truth and leave. Le Guin's story seems to make the point that our delight, safety and abundance are intimately tied with the suffering of others somewhere else. Yet at the same time, this parable invites us to recognize that even those settings that, we believe, are peaceful or devoid of the effects of conflict often owe their very conditions of existence to some form of violence. Cruelty and coercion, Le Guin appears to be telling us, almost always lurk behind harmony and consensus.[1]

In this chapter, I turn my attention to the act of political violence that led to the murder, displacement, and dispossession of Armenians who once lived in the Ottoman Anatolia, which has drastically shaped the social and economic life in Turkey. I explore how the disappearance of Armenian lives, and the destruction and plundering of their property more than a century ago continues to reverberate in the present. Even though the genocide remains officially denied by the Turkish state and misrecognized or disremembered by the majority of its citizens, Yusufeli's residents, I argue, become exposed to the effects of its legacy as they wait for their coming annihilation and displacement. This exposure is most commonly manifested through the curious deployment of violent processes (war, annihilation) as metaphors to describe the physical damaging of the valley in its different temporal registers. On other occasions, earlier episodes of violence and displacement are explicitly invoked, and their victims are remembered to place the suffering of the past in juxtaposition with grievances in the present. The dead of the past, in other words, reappear in discourse precisely at the moment when life in the present, at least in the form and sense that the living know it, is about to extinguish.

This is not haunting. Strictly speaking, it does not serve a ferocious recognition of historical crimes, either. And unlike in Omelas, it does not make anyone walk away from the town. What it nevertheless does at times in Yusufeli is to render the local community susceptible to the ruins that history has accumulated in the past hundred years, opening transitory cracks in its official narratives of triumph and victory to which its members are invariably interpellated.

Why do the past figures of the genocide resurface amid the ongoing ruination set in motion by capitalist development? I have argued in the last chapter that the destruction of the valley and the displacement of its residents become the context in which an entrepreneurial drive took shape to galvanize various forward-looking strategies of accumulation and commodification. Here, I want to add that despite that, the majority of residents remain emotionally attached to the disappearing built environment, nature and ways of living, and visibly suffer when they encounter instances of this transformation in their everyday lives. It is the absence of a political agency that could prevent this loss and injury, I speculate, that instigates another temporal orientation among the residents that I will call "looking backward." Timid murmurs, pale marks, and fleeting sensibilities that I register in this part of my ethnography highlight the persistence and continuity of violence, its capacity to connect and transcend time and space, and its tendency to transmogrify in its transfers between human, natural, and built environment.

Echoes of past violence in the contemporary moment may have the power, to go back to Gramsci's terminology, to unleash portions of the critical thought embedded in the common sense enveloping the valley's Turkish-Sunni identity, even if it never quite develops beyond its embryonic form. Yet again, the same violence still reproduces this community to this day as its material effects such as dispossession is not a "one-time event" but exist "as a constantly concurrent combination of capitalism's three moments and temporalities of origin, development and crisis" (Tomba 2013, mentioned in Harootunian 2019: 89). I therefore fix my ethnographic gaze on how the theft of Armenian property, which in part motivated the large-scale extermination, continues to play a role in the processes of accumulation a hundred years later. Through the use of historical records, archival material, a survivor memoir, and several ethnographic encounters, I trace in this chapter how the state-enforced transfer of Armenian properties to Turkish-Sunni residents and the ceaseless search for "Armenian treasures" in Yusufeli continue to have an impact on the wealth and property accumulated or anticipated to be accumulated in the present. By looking at the reverberations of the originary violence across time and space, I show how, in other words, the originary accumulation is once again produced under the bulldozer variant of capitalist development.

Thinking through the rubble of bulldozer capitalism in the mirror of past ruins, as some of my interlocutors do here, reveals the necessity to adopt an analytical gaze that eventually does not limit itself to contemporary projects of resource extraction, but that tries to account for the layeredness of expropriation and accumulation and what this politically and conceptually implies for extinction and survival. In some sense, my material on Yusufeli echoes, once again, Anna Tsing's (2015) argument that life continues to flourish in capitalist ruins, paradoxically, thanks to particular patterns of ruination tied to historical legacies of valuation. However, in place of outright embracing the collaborative survival of multispecies as the ontological model behind history making, in this chapter I am inclined to linger a bit longer on what has already disappeared and how it informs what is about to become extinct. Perhaps, this offers a glimpse of capitalism's present and future spaces of contradiction since they are constituted, as Walter Benjamin (2002) seems to be saying in *The Arcades Project*, not only by what endures, remains or is salvaged, but also by what has irrevocably vanished. The debris of history in the middle part of the Çoruh Valley does not lie there motionless. Bridging timeliness to speculate on the death of nature and the built environment animates the past figures of sovereign violence.

Violence and the Originary Violence

From the residents' perspective, the local bullfighting festivals organized every year between April and September are the main theaters of conflict where the pursuit of dominance and social status plays out over the spectacle of injury and blood. Echoing at least one anthropologist's observation on the sharp difference between tribal communities' social relations and cosmologies—the more peaceful and egalitarian the former is, the more conflictual the latter becomes (Graeber 2004)—the residents themselves often reflected on how the overall lack of violence in their everyday lives contrasts with their enthusiasm for watching and taking part as bull owners in the fights between animals. As a matter of fact, even these fights are usually without any spectacular violence. Standing still in a rectangular sand field, two bulls often gaze at each other uninterestedly for minutes as their owners gently prod them with long wooden sticks. Hundreds of spectators flock to the arena with their teapots and finger food early in the morning and with their eyes glued to the encounter between the animals, wait for the moment when one of them eventually begins

Figure 5.1. Bullfighting in Derekapı, Yusufeli, 2013. © Erdem Evren.

chasing the other. Those rare moments when one animal becomes agitated enough to directly attack its opponent and pierce its body with its horns are recounted again and again for the following days and weeks. For those like me who are less interested in bulls, there is always the opportunity to try our luck with the DIY roulette games in the hope of winning a pack of cigarette or to drink beer semi-secretly with others behind the tall bushes.

More crucially, however, these festivals were sites where not only is the local web of meanings enacted, but also capital and power accrued during the construction boom are exhibited and rituals for the nation are staged. Previously funded independently by each village council, the events that I attended after 2013 increasingly revealed the marks of shifting class composition and the political underpinnings of the newfound wealth in the town. Banners attached to the sides of the fighting arena displayed the names of the festival sponsors: either representatives of construction companies building dams and roads in the valley or the shop owners who became more affluent thanks to the real estate economy. Each year, the person making the biggest donation to the event, without exception one of the local contractors, was given the title of master of ceremony (*güreş ağası*) and had the opportunity to rub shoulders with the visiting MPs and governors for the day. Between the fights, the presenters addressed the crowd with a microphone to demand prayers and a minute of silence for the martyrs who fell while defending the country. Occasionally, they also publicly honored the relatives of the conscripts from the town who were killed fighting in the Turkish state's ongoing war against the PKK and performed long soliloquies on the sacredness of the nation and the flag and the unity and indivisibility of the country. "We shall not hesitate to defend our fatherland against domestic and foreign threats," we were often told, "just like our grandfathers and great-grandfathers have done before us."

In its everyday incarnations, this Turkish-Sunni identity provided the grammar through which attachments to the valley and the responses to the changes in the physical environment are expressed. What was going to happen to the town's mosques and cemeteries after the submergence of Yusufeli was a common worry repeatedly raised by my interlocutors, despite the planners' assurances throughout the years that they, like in other parts of the valley, would be carried to the new settlement area. A close interlocutor until 2015 and the brother of my imam friend whom I mentioned in Chapter 1, Acar Bey, on the other hand, often lamented that he would never have the opportunity to celebrate Bayram with his extended family in their village, Irmakyanı, once the Yusufeli Dam project is completed. He also complained that the ongoing destruction of the valley and the coming submergence were eroding his sense of belonging to his lived environment. As an ambitious civil servant specialized in writing funding applications for international donors such as the European Union for various small development projects in the town, Acar Bey eventually ran as an independent mayoral candidate in local elections in 2014, promising to negotiate with the state compensations not only for economic grievances but also for "social losses" (*toplumsal kayıplar*). In the months leading to his candidacy, I would often find him in his office with teary eyes, listening to melancholic songs on the radio. His suffering was exemplary of an excess even the widespread expectations to profit from ruins and ruination cannot entirely obliterate. Shortly after the elections, he asked for his transfer to a town in western Turkey and quietly left without bidding farewell to his fellow townsfolk.

The local anti-dam/anti-displacement campaign flourished at the intersections of the militant and civic manifestations of this Turkish-Sunni nationalism. As I already mentioned in the introduction to this book, all three prominent members of the local cultural association,

Muhsin Bey, Ragıp Bey, and Hikmet Bey, who launched and coordinated the opposition against the Yusufeli Dam project and other hydropower projects in the middle part of the Çoruh Valley, were sympathizers or according to rumors even perpetrators of the anti-Kurdish and anti-leftist state and rightwing paramilitary violence. Politically sidelined for their strong anti-AKP convictions during the party's meteoric rise in the town, as in other parts of the country, these activists increasingly resorted to this nationalist-conservative rhetoric in order to animate the earlier mobilization against the dam project. On the one hand, they portrayed the dam project as part of a sinister plan that Turkey's enemies had concocted and the AKP was willingly implementing to divide the country along ethnic lines. On the other hand, they tried to appeal to the residents' nationalist sentiments by arguing that the displacement of a community that had always been loyal to the Turkish state was a grave betrayal. To prove this point, they gave me and anyone else who was interested long genealogical accounts embellished with stories from a series of wars from the Ottoman-Turkish history in which their grandfathers and great-grandfathers ostensibly took part.

I was therefore bemused when a friendly employee from the local land registry office where I was trying to conduct interviews in 2014 asked me one day if I wanted to see the records of a population exchange that took place between Turkish and Georgian civilians in the 1920s. I knew about the Crimean War of 1853, the Russo-Turkish War of 1877–1878, and the Russian Empire's subsequent occupation of Kars and Batumi provinces (the latter of which Artvin was a part), followed by the "skirmishes" between the Ottoman army and Armenian forces during World War I, which have all left a mark in the region, but I had absolutely no clue what she was talking about. A few days later, she took me to a room the size of a broom closet in the basement of the government office to show me a mountain of notebooks written in Ottoman Turkish, which was thus regrettably inaccessible to me. A few of them in Latin script, however, contained records offering a tiny glimpse of the lives of these residents who once lived in the area. Some of the documents that I quickly noted down read as follows:

> Field in Kiskim, which was passed from Vartan, son of Pertek, to the treasury, is sold in auction in 1926.
>
> Rice field in Erkenis, which was passed from Kirkor, son of Kazar, to the treasury, is sold in auction to Ahmet Bumin.
>
> The entire meadow (832 m^2) was passed from Ahamis and Kirkor to the treasury and is sold in auction in 1926.[2]

One thing that immediately caught my eye was that the year 1926 appeared in almost all documents as the date when the treasury auctions took place, and the first installments were paid. The new owners of these rice fields, meadows, and gardens, who were often civil servants then working at the land registry office, continued to make regular payments until as late as the 1940s, which explains why these records were kept in Latin script. Something else that became clear, however, was that the "Georgians" whom my interlocutor said were the object of a population exchange in the 1920s were none other than the Armenians who either fled their homes or lost their lives before and during their deportation from the Ottoman Empire in 1915.

Despite increasing attention to the geographical variations in the execution of the deportation orders by the Young Turk regime, few comprehensive local histories have so far been produced on the fate of the Armenians from the Çoruh region.[3] One reason behind this gap in scholarship is the relatively small size of the Armenian population as a result of their mass migration to Caucasia in the early nineteenth century. Nevertheless, in one well-documented case, about ten thousand Armenians from Hodorcur, a large settlement that lies to the southwest of Yusufeli and is today known as Sırakonaklar, were exterminated in 1915. Equally importantly, the long and complicated process by which the "abandoned" Armenian property was sequestered on a massive scale remained missing from most historical studies on the genocide until very recently. Determining the local specificities of the property transfer remains an arduous task because the land registers from the period, which are stored at the Land Register General Directorate in Ankara and which several historians presume to contain the most detailed account books of confiscated property, are still closed to researchers. Those records that I serendipitously stumbled upon in the basement archive corroborate the findings of two recent studies that argue that a major decision made by the government on 13 June 1926 nearly finalized the judicial framework for the seizure of the "abandoned" properties by the Turkish state. It therefore removed the latest barrier to transferring the property to Turks as part of the goal of establishing a Turkish national economy set up by the Committee for the Union and Progress (CUP) and continued by the new Republican Government after 1923 (Akçam and Kurt 2012; Üngör and Polatel 2011).[4] More importantly for my purposes here, these documents also give us reason enough to believe that at least some of the land in and around Yusufeli that the current residents dwell in, build upon, and anticipate converting into cash in the form of compensation payments once belonged to the Armenians who lost their homes and lives. As a matter of fact, in our subsequent conversations, my interlocutor from the land registry office revealed that much public property, including those parcels of land where the town's school and hospital are built, also once belonged to the "Georgians."

Reverberations

In his recently translated memoirs, Raffaele Gianighian (2016), one of the seven people from Hodorcur who survived the genocide, sheds some more light on the fate of the Armenian property that was left behind. Returning to the region as an old man in 1977 from his adopted country of Italy "to search for his fatherland," Gianighian finds his village in ruins. As he strolls through the streets, he looks at the rubble of the church, the cemetery that bears the marks of destruction, and the mournful faces of old Muslim residents who personally knew his family. A woman from the village who volunteers to accompany him to his village Kisak begs him to tell her where the "Armenian treasures" are hidden, crying that she cannot bear living in this cursed place anymore and needs the money to move to the United States. In the evening, the youngsters from the village ask him what happened to the old villagers of Hodorcur. Gianighian recounts the murder of his fellow villagers at the hands of soldiers, and Turkish and Kurdish civilians; the death of several relatives because of illnesses as they are forced to move from one place to another; and his life in his final destination, Büyükbağ, where he starts working as a goldsmith for the local Kurdish feudal lord, who eventually forces him to convert to Islam and gives him the name Abdullah.

Almost forty years after Gianighian's return to the region and nearly a hundred years after the genocide, the tragedy in Hodorcur was still a conversation topic to which young men from Yusufeli occasionally turned. One day, in one of the new cafés in the town center, a large group of friends suddenly started to talk about what happened to the Armenians from the region. One interlocutor known for his interest in the history of the region explained to others that all the Armenians from Hodorcur were murdered and their property confiscated. His audience became visibly furious with the "ancestors" for committing this crime and began to curse their grandfathers and great-grandfathers. Yet, a few of them asked whether there were any Armenians treasures to be discovered.

Even though I did not attribute much significance to this conversation at the time, I was still slightly surprised by my interlocutors' reactions, given the Turkish state's power to consolidate its official denialist narrative of the Armenian Genocide among its contemporary citizens in schools and the media. To this day, this narrative relativizes the large-scale annihilation of the Armenian community and the responsibility of the Ottoman-Turkish administrators and civilians by alternatively framing the event as a "mutual slaughter" (*mukatele*), attributing the majority of deaths to "unavoidable" illnesses that spread during the deportation, or even disputing the veracity of Armenian deaths by seeking to create a remembrance culture around the Turkish civilians who lost their lives at the hands of Armenian gangs.

This latter trope, I observed, was particularly powerful in the ways in which the region's Armenians are portrayed. Closely following the writings of nationalist historians such as Fahrettin Kırzıoğlu, the long text in the governorate's (*valilik*) website devoted to recounting the history of Artvin represents the Turkish gangs' activities as a purely defensive measure against the killings of civilians by Armenians.[5] The texts and booklets produced under the auspices of the mayor's office similarly stress that Armenians' collaboration with the Tsarist forces led to the murder of several Turkish civilians. Several interlocutors invoked this argument in our conversations when they randomly mentioned parts of their family histories concerning the fight against the Armenian gangs or stated the name of a great-grandfather who became a "martyr" defending his village. A recent book by historian Candan Badem (2018) on the Tsarist military-communal administration in the cities of Kars, Ardahan, and Artvin between 1878 and 1918, however, not only provides useful corrections to the nationalist historiography of the region but also reveals the extent of the brutality employed by the valley's Turkish-Sunni residents against the Catholic Armenians from Ardanuç, a town located ninety kilometers east of the current Yusufeli town center. Both the survivor accounts in the Russian and Georgian archives and the transcriptions of a high school oral history project conducted with the elderly in Ardanuç in the late 1980s that Badem accessed leaves no doubt that it was the gangs coming especially from Yusufeli and the village of Hod (Maden), now part of Artvin, that took part in the extermination of Armenian lives. Similar to countless other places in Anatolia, several villagers and former prisoners were recruited and weaponized by the agents of the Special Organization (*Teşkilat-ı Mahsusa*), the special forces unit established by the CUP administration, and took part in the killing of the majority of adult males, plundering of their property, raping of their wives and daughters, and the drawing of lists for taking the surviving women as concubines. While several Armenians were killed in Ardanuç and the surrounding settlements, the majority seems to have been thrown alive into a nearby canyon called "Cehennem Deresi" (Hell's Creek). While some of the Turkish families were indeed killed by Armenian gangs, none of the violence against Armenian civilians is evoked by the majority of current residents.

I was therefore genuinely baffled to encounter the figure of the dead Armenian reappearing in other conversations with older and markedly more rightwing residents in the context of the ongoing infrastructural violence. One day, Hikmet Bey, the retired police chief who had played an important role in organizing the opposition against the dam projects in the region for years, was showing me the landscape photos that he recently took with his new digital camera in the valley. Initially cheerful about sharing the products of his new hobby, his face suddenly turned grim as if he just recognized something unpleasant. He told me that even though his grandfather also fought against the Armenians, their old enemies had never violated nature and its beauties. "The Armenian" understood and respected the flowers, rice fields, olive groves, and the river itself, whereas the new enemy, that is, the government and the companies, only intend to obliterate them. This is why, he reflected, their violence is even more destructive. Another resident of Yusufeli, Hasan Bey, who had recently moved to Istanbul at the time I conducted an interview with him, similarly invoked parallels between the past and the present:

> This [the displacement of people] is a matter of identity. The people still living in Yusufeli will find themselves in a situation similar to what we, those who have already left, are now going through. Once you lose your hometown [*memleket*], you begin to pursue the traces of your identity. Like Armenians who have been doing it since their deportation. How similar these two are, I mean, the deportation [*tehcir*] from Yusufeli [that the current residents will endure] and the deportation that Armenians became subject to. Of course, the circumstances were different then. We won't have a place to go back to if we lose our jobs in Istanbul or in Antalya. Those Armenians [who used to live in Anatolia] now living in France don't have a place to go back. Their land has already been confiscated. Maybe one difference is that in our case there won't even be an archive that registers our presence.

Hasan Bey's comparison between the future plight of his fellow townsmen and women and the fate of the Ottoman Armenians certainly has its limits and problems, as he himself self-consciously notices. What Hikmet Bey's and his narratives nevertheless seem to suggest is that, as much as the hegemonic nationalist inculcation through which certain lives are glorified and celebrated and certain deaths are glossed over remains a very powerful process, it is never entirely complete. It is possible for even some of the loyal adherents of this ideology and the perpetrators of its violence to momentarily digress from established scripts to constitute different relations of victimhood and culpability. To repeat, this by no means serves to facilitate the beginnings of a proper confrontation with past atrocities and their present repercussions as an ownership regime based on the theft of property. Nor would it prevent, as we saw in the last chapter in the context of a lynching attempt against Kurdish construction workers, the repetition of this crime in the contemporary moment under different circumstances. This is still a transitional sensibility or an affect, reminiscent of what Yael Navaro-Yashin (2012) studied in her ethnography on Northern Cyprus as an eeriness or disturbance formed at the interstices of nature, the built environment, and human bodies.

This affect emerged, I argue, in a context in which the residents began to wait for the coming destruction of their surroundings and habitual ways of living after they lost the means to contest it. Regardless of whether they actively partake in speculating on this destruction by bringing to life strategies and visions of accumulation or, on the other hand, continue futilely trying to stop it from materializing, they recognize that they cannot avoid the fate of loss, displacement, and alienation. This is perhaps why the dead of the past occasionally reappear to them as the harbinger of their own unfortunate inability to muster their will to remain viable in the future.[6]

Recall the scene, which I described in the introduction to this book, where my driver Orhan and I caught from a hilltop on the second day of Bayram the sight of ruination going on in the village Sirya/Zeytinlik. Remember also that what we saw produced much unexpected discomfort for him. This is perhaps because for him the immediate, personal, and accelerated nature of self-imposed destruction juxtaposed so powerfully with the slow and protracted devastation that he is used to experiencing in the valley that the logic of valorization failed, momentarily, to compensate for the ongoing ruination. Or put another way, "looking forward" collapsed in that instance into its opposite, that is, "looking backward."

Hunting for Treasures

My accidental discovery in the archives no doubt made me see Yusufeli under a different light. Every time I took strolls in its meadows or heard a villager explain that his vegetable garden used to be a rice field, I could not help but wonder if this was yet another piece of confiscated Armenian property that its current possessors were waiting to cash in after the expropriation. It was only through the very end of my research that I came to realize that not only the buildings and gardens but also the "underground" was a site where the past violence is being tried to be converted into possible value (von Bieberstein 2017). What captivated some of Raffaele Gianighian's fellow villagers already in the 1970s and what some of my young friends timidly fantasized about pursuing during the time that I spent in the town was indeed an ongoing pastime activity for several men from Yusufeli: treasure hunting or *definecilik*.

Treasure hunting, to paraphrase one seasoned practitioner based in western Turkey who earns a living by selling state-of-the-art metal detectors and regularly posting videos on YouTube to inform fellow hunters or *defineci*s, is essentially the search, with the permission of the state, for the money and valuable objects left by minorities a hundred years ago.[7] Even though they have certainly not been subject to a genocide, as the same YouTuber explains in his videos following the denialist script, finding the possessions of Ottoman Armenians left behind before their deportation from Anatolia is the main objective of this activity. Often hidden inside wells, old houses, churches, and shrines across the rural parts of Anatolia, one first needs to learn how to decipher certain marks such as letters, crosses, and animal figures carved or painted on rocks or walls or to find and read the maps left by former owners to be able to discover these troves.

What a particular symbol discovered in the terrain might possibly denote, which would bring the *defineci* closer to finding the location of "Armenian treasure," prompts thousands of men from different parts of the country to turn to the Internet every day. In online treasure hunting forums and more recently the comment sections of YouTube videos, they consult older and more experienced hunters, often referred to as *üstad*s or masters, by posting photos of maps and signs that they discovered in areas close to their towns and villages. I was touched in particular by one photo posted in an online forum, showing the interior of an old village house seemingly left completely intact even after more than a century. A wooden table, rusty utensils, and Christian icons still adorning the tiny kitchen gave the impression that its photographer was possibly the first person to enter this place since its owners were either deported or killed. This man now wanted to know how he could seize the hidden Armenian gold.

But then, even correctly reading the symbols that point to the location of golds and old possessions may not be enough for unearthing them. Armenian treasures are believed by most of its hunters to be protected by spells (*tılsım*), which either create illusions to confuse the *defineci* or put his life in jeopardy. Ghosts or ghost-like figures like djinns (*cinler*), the supernatural beings in Islamic mythology, often appear from nowhere in houses believed to host Armenian possessions or during the excavation of the land. It is therefore not uncommon that an imam or someone with some religious knowledge accompanies the diggers to break the spell and fend off or neutralize the power of these spirits protecting the treasures by reciting prayers.

Remarkably, such practices, narratives and imaginaries that mediate everyday engagements with treasure hunting have received attention in the ethnographic literature on Turkey only in the past couple of years. Based on his research among a Greek-speaking yet ardently Turkish nationalist community living in a remote valley in Trabzon, northeast Turkey, Erol Sağlam (2020) writes of his interlocutors' search for Greek or Armenian treasures in seemingly natural places such as lakes, mountains, and hilltops. According to this account, treasure hunting is not a destructive activity conducted in pursuit of enrichment. Denying its own ethnic background despite continuing to privately speak an archaic dialect of Greek, this community's dealing with the valley's abject cultural heritage rather generates a nondiscursive, affective, and topographical relation with a past that it has been forced to forgo after the construction of a monolithic Turkish national identity. Özlem Biner's (2020) ethnographic research in the multiethnic/religious city of Mardin in eastern Turkey, at a time when the clashes between the Turkish state and the PKK temporarily ceased, looks at the entanglements between the booming heritage economy and the hunt for treasures. Decrepit stone house mansions left by Armenians and other ethnic and religious minorities in the center constitute part of the city's ruined heritage that gained a market value around this time. Their new owners, often middle- and lower-class Kurds, obsessively dig on their properties to be able to unearth the buried money and possessions that they believe were left by Armenians. Biner considers the activity of digging in these ruined buildings generative of ordinary affects that relate to encounters between and among human and more-than-human entities such as djinns, through dreaming, apparition, haunting, and hoarding. These affects point to desires for recognition and enrichment, as well as the worries about the future, in this geography, which for decades has witnessed state violence and intercommunal tensions. Finally,

Alice von Bieberstein's (2017) study situates treasure hunting in the city of Muş in Turkey's Kurdistan at the intersections of the foundational violence of the nation-state and the commodity economy of the modern capitalist system. Deeply committed to the Kurdish political movement's struggle against the Turkish state and thereby embracing the PKK's official position that some parts of the Kurdish population have been complicit in the extermination of the region's Armenians, her interlocutors still passionately hunt for Armenian gold. Von Bieberstein explains that this has to do with the fact that these objects become fetishes, untethered from the violent histories that engendered them, to accommodate what Michael Taussig (1999: 5) calls "the public secret"; "that which is generally known but cannot be articulated." *Definecilik*, as a form of looting and spoliation, she argues, therefore occasions this community's participation in the denial of the originary violence and dispossession that founded the Turkish polity, despite the Kurdish movement's recognition of the genocide and expression of solidarity with all oppressed groups.

I learned about my Turkish-Sunni interlocutors' involvement in the afterlives of this originary violence and its economy of looting belatedly and indirectly when my friend Okan, who earlier mentioned the massacre of Armenian civilians from Hodorçur/Sırakonaklar, spoke one day in summer 2017 about the most passionate treasure hunters in Yusufeli. Apparently, for several years, Muhsin Abi has spent much time and money to search for the region's "Armenian treasures." Accompanied by a few other shop owners, he reportedly went on "hunting expeditions" usually late at night, following the clues left in old maps or digging underneath those rocks on which figures and symbols are carved. Not unusual for most hunters, his efforts after all these years remained mostly fruitless, however, as he ended up unearthing only a few coins that Okan dismissed as *teneke* (literally, tin) or useless rather than the desired gold. In a convoluted story narrated to me by another interlocutor, a few locals collaborated with a German citizen who showed up in the town one day with an old map that confirms the location of a trove near the town center. While these hunters managed to find some valuable coins and artifacts, they apparently got arrested by the police when the German man tried to take his share of the loot out of the country.

One family living in the town center, according to many interlocutors, certainly owed its wealth to "Armenian treasures" that they found in a village more than two decades ago. Cashing in the large trove, its members opened a small shop on the main street, which evolved over the years into a much larger and fancier one specializing in selling household equipment manufactured by well-known domestic and international companies. During the real estate boom in Yusufeli, this family, like other affluent or well-connected shop owners, erected modern apartment buildings in the town center, expanding their wealth even more by attending to the new construction and rent economies.

Thus, treasure hunting makes visible that violent dispossession based on the large-scale expropriation of wealth is not a concluded event but a foundational condition of the "capital-nation-state" assemblage that continues to give shape and direction to capitalist development. In its afterlife, this violence reproduces the existing communities of Anatolia not only through its ghosts and ruins resurfacing under the new circumstances occasioned by the transformation of both urban and rural areas, but also because of the ongoing looting and exchange of property and belongings left behind by its victims. Reverberations of the Armenian Genocide generate ordinary affects such as eeriness and disturbance also among Yusufeli's Sunni-Turkish community, as I argued above, to mark their fears about the destruction and displacement caused by dam projects. Yet, economies of plundering reveal the limits of these affects' power to unsettle accounts of denialism and to enable the recognition of past atrocities. If the ceaseless optimism embedded in drawing deeper meanings from signs and figures to acquire guidance for potential enrichment turns d*efinecilik* essentially into a "speculative and future-oriented endeavor," as von Bieberstein (2017: 182) argues, it instead correlates with and attends to residents' visions of and projects for tapping into ruins and ruination.

I studied in this chapter how the extermination of Armenians from the Artvin region and the expropriation of their wealth become at once the excess of and the conduit for the ongoing efforts to speculate on infrastructural violence in the middle part of the Çoruh Valley. I made the point that the residents' gradual accommodation to this violence, and the social and political relations that facilitate the formation of a consensual practice that interrupts their anti-dam politics in this resource frontier should be considered also in connection with the originary violence that unleashed the originary accumulation. Be it in the form of making a profit out of the properties expropriated in the past or looting the possessions of former residents, the efforts to seek enrichment amid the ongoing destruction make the genocide and its originary accumulation resurface in the present.

Notes

1. In my earlier work (Evren 2012), I used Le Guin's story to reflect on the conscientious objectors' movement in Turkey. For a work by an anthropologist that delves even deeper into this parable to make sense of the governance of difference under late liberalism, see Povinelli 2011.
2. As I mentioned earlier, Kiskim is the old name of Yusufeli. It was a district administered by the city of Erzurum for several decades. A large village named Ersis, today known as Kılıçkaya, was the center of Kiskim until the mid-1920s.
3. For a historical ethnography that discusses the local dynamics of the Armenian genocide in Van, see Türkyılmaz (2011); for a study on Adana, see Üngör and Polatel (2011). For two works on the deportation and murder of Armenians from this region, see Aksu (2013) and Kévorkian (2015: 405–31).
4. It is not my intention here to enter into a debate over the complicated legal process by which the transfer of Armenian property was completed, but the material that I came across seems to suggest that the Treasury auctions and the distribution of title deeds to new owners began to take place two years earlier than generally assumed by historians. Both studies of mass sequestration that I cited earlier seem to argue that, as important as it was, the ordinance in 1926 still had a caveat, because it stipulated that the "abandoned" properties recognized by the state after this date still technically belonged to Armenians. By contrast, a decree passed in 1928 clearly states that the immovable property liquidated by the Treasury will not be given back to its former owners, prompting scholars to believe that property transfer officially started after this year. Yet, the documents from Yusufeli clearly indicate that the first installments for the Armenian properties were paid in 1926.
5. For the text on the history of Artvin in the governorate's website, see http://www.artvin.gov.tr/artvin-ili-tarihcesi.
6. For a similar argument, see von Bieberstein (2021).
7. Legally speaking, treasure hunting seems to fall into a gray area. Although it is stated in the regulations that treasure hunters may receive cash payment in exchange for parts of the loot if they apply to the state in advance for receiving licenses for digging, very few people actually do that. In almost all cases, local authorities turn a blind eye to these activities. For a video where the aforementioned *defineci* talks also about the legal basis of hunting "Armenian treasures," see www.youtube.com/watch?v=0LIdZafL7y0.

Conclusion

In the morning of 1 April 2019, Yusufeli's name appeared on the pages of virtually every single national newspaper in the country. This time news reports were not covering the latest developments about the Yusufeli Dam project but the results of the municipal elections held the previous day. Eyüp Bey, who once again ran as a candidate from the AKP had won by only four votes. After the recount, three more ballots were announced to be spoiled to bring down his lead on the CHP candidate to a single vote. Following subsequent legal challenges by the CHP, elections were repeated on 2 June and this time Eyüp Bey won the elections with a margin of a few hundred votes.

The AKP's very near loss in Yusufeli, together with its defeat in Artvin and a few other towns of northeastern Turkey, received much commentary from locals, as well as the journalists and academics writing on this region. For some, this result should be considered as the outcome of a successful country-wide electoral strategy formed between the CHP and the IYI Party, and tacitly supported by the HDP, which also helped to win back several large cities including Istanbul and Ankara. Others drew attention to the unique case of the town to ask if its residents chose to belatedly punish the party-state for its fallen promises. In either case, the expropriation of all land and property in Yusufeli and its villages, as well as the construction of much of the new settlement area and the Yusufeli Dam, were completed by early 2020. Even though the DSI announced around the same time that the resettlement process is expected to begin by March 2021, this has not happened yet. Photos recently taken in the town and shared in online forums show that only a few buildings have been evacuated by their owners. In the meantime, despite a devastating economic crisis, becoming even more crushing with exchange rate shocks hitting the country almost every other week and recently imposed lockdown measures because of the COVID-19 pandemic, Turkey's bulldozer capitalism continues to expand into new frontiers. In addition to several ongoing construction and extraction projects, which continue to be met by local protests on an almost daily basis, the plans for mining in five hundred new sites across Turkey were recently made public.

Bulldozer Capitalism studied the rise and decline of an anti-dam/anti-displacement campaign and the political responses to other extractive projects that it helped to shape in its aftermath. It showed that people can be made to accommodate their own dispossession and displacement if they are directed to negotiate, invest in, and speculate on the destruction of their built environment and nature, and their material and immaterial bonds, wealth and activities. This happened, I argued, essentially after their spatial and temporal sensibilities and experiences of their valley's material transformation were starting to be controlled and managed through the interventions and arrangements of party networks across the local and national scales. The party-state's incitements for the residents to make calculations about the effects of infrastructural devastation, I claimed, are ultimately constitutive of an entrepreneurial subjectivity strongly predicated on ways of thinking and acting connected with the tools and mechanisms of finance. In this book, I discussed the efforts that went into the cultivation and direction of entrepreneurial subjects as part of a form of governing without resort to violence.

This last point can be read as a contribution to an unending political and academic debate over the Justice and Development Party's success in ruling Turkey uninterrupted for the past nineteen years. By contrast with those explanations that reduce the AKP's power alternatively to sheer physical force, politics of religion, poor relief strategies, or the false consciousness of its support basis, my ethnography traced the conditions that made possible the formation of a new form of consensual political practice at the intersections of economy, identity, and infrastructure. The fluctuations and contradictions of global capitalism in the twenty-first century, as well as the drastic shifts witnessed in the planning and finance of development, to which the campaign in Yusufeli also contributed by articulating itself into a transnational space of anti-dam activism, had the effect of strengthening the links between the national and local scales. The party-state owes its power to act upon the expectations, fears, and desires of the residents in the middle part of the Çoruh Valley to a plethora of actors who mediate these links in order to facilitate, or at least promise, interventions in a number of critical social, material, and (para)legal matters. This hegemonic project is not always without its own contradictions and limits as different interests may clash or promises quickly fail to materialize or to generate discontent. But in this part of Turkey, for a long time it remained effective enough to make people believe that dispossession and displacement could be tolerable if they could bring to life their own projects of speculative accumulation and commodification based on the protracted ruination of the valley. But beyond its possible contribution to discussions on the politics of Turkey, a number of other insights that come out of *Bulldozer Capitalism* may potentially speak to other contexts. I will end this book by briefly mentioning two of them.

First, the formation and interruption of politics that I traced in the middle part of the Çoruh Valley ultimately shows how consent is generated amid the experiences of crisis. The crisis in question in this resource frontier is simultaneously social, economic, and ecological, like in so many other places in the contemporary moment. This book can therefore also be read as an inquiry about why ordinary people actively take part in the destruction of their own livelihoods and the lives of others—human and nonhuman. This question continues to occupy countless commentators and observers from the US to Hungary, India to the Philippines. In this book, I tried to go beyond the "negative methodology" (Navaro 2020) that characterizes several political ethnographies that came out in recent years and, rather, to capture, through the image of bulldozer, the dialectic relation between the negative and the affirmative that comes to constitute the political.

Second, my ethnography tells us that being inside the capitalist system also means being inside the logics and tensions of sovereignty. Genealogies of sovereign violence remain critically important for the form and direction that contemporary processes of accumulation take. At the same time, the extraction of value from land and natural resources does not always depend on the destructive effects achieved through the suspension of law. Neither does it necessarily require the removal of legal protections by the sovereign power. One of the findings of this book is that hegemony, understood as the consensual political practice, is ultimately a productive effect of sovereign exceptions. This point may be worth considering while rethinking the links between capital and sovereignty, as well understanding how authoritarian political projects from across the world continue to consolidate their power through extractive projects.

REFERENCES

Abram, Simone, and Gisa Weszkalnys. 2011. "Introduction: Anthropologies of Planning—Temporality, Imagination, and Ethnography." *Focaal* 61: 3–18.

Adalet, Begüm. 2018. *Hotels and Highways: The Construction of Modernization Theory in Cold War Turkey*. Stanford: Stanford University Press.

Adaman, Fikret, Bengi Akbulut, Yahya Madra, and Şevket Pamuk. 2014. "Hitting the Wall: Erdoğan's Construction-Based, Finance-Led Growth Regime." *The Middle East in London* 10(3): 7–8.

Akbulut, Bengi, Fikret Adaman, and Murat Arsel. 2018. "Troubled Waters of Hegemony: Consent and Contestation in Turkey's Hydropower Landscapes." In *Water, Technology and the Nation-State*, edited by Filippo Menga and Erik Swyngedouw, 96–114. London: Routledge.

Akça, İsmet, Ahmet Bekmen, and Barış Alp Özden. 2014. *Turkey Reframed: Constituting Neoliberal Hegemony*. London: Pluto.

Akçam, Taner, and Ümit Kurt. 2012. *Kanunların Ruhu: Emval-i Metruke Kanunlarında Soykırımın İzini Sürmek*. Istanbul: İletişim.

Akçay, Ümit. 2018. "Neoliberal Populism in Turkey and Its Crisis." *IPE Berlin Working Paper* No: 100.

Akçay, Ümit, and Ali Rıza Güngen. 2019. "The Making of Turkey's 2018–2019 Economic Crisis." IPE Berlin Working Papers, 120. Berlin School of Economics and Law, Institute for International Political Economy (IPE).

Aksu, Cemil. 2013. "Artvin Ermenilerine Ne Oldu?" *Tarih ve Toplum: Yeni Yaklaşımlar* 16: 91–126.

———. 2016. "Derelerin Özgürlüğünden Yaşamın Özgürleştirilmesine: HES Karşıtı Kitle Hareketlerinin Politik Ufku Üzerine Bir Deneme." In *Sudan Sebepler: Türkiye'de Neoliberal Su-Enerji Politikaları ve Direnişler*, edited by Cemil Aksu, Sinan Erensü, and Erdem Evren, 395–427. Istanbul: İletişim.

Allan, Matthew. 2011. "Mediterranean Shareholders Force Management Shake-up." *The Northern Miner* 97(7). Retrieved 29 December 2020 from https://www.northernminer.com/news/mediterranean-shareholders-force-management-shake-up-2/1000406674/.

Amalric, Franck. 2005. "The Equator Principles: A Step Towards Sustainability?" CCRS Working Paper Series No. 01/05. Center for Corporate Responsibility and Sustainability, University of Zurich.

Appadurai, Arjun. 1996. *Modernity at Large: Cultural Dimensions of Globalization*. Minneapolis: University of Minnesota Press.

Arboleda, Martin. 2020. *Planetary Mine: Territories of Extraction under Late Capitalism*. London: Verso.

Arıcan, Alize. 2020. "Behind the Scaffolding: Manipulations of Time, Delays and Power in Tarlabaşı, Istanbul." *City and Society* 32(3): 482–507.

Arsel, Murat. 2016a. "Reflexive Developmentalism? Toward an Environmental Critique of Modernization." In *Environmentalism in Turkey: Between Democracy and Development*, edited by Fikret Adaman and Murat Arsel, 29–48. Aldershot: Ashgate.

———. 2016b. "The Bergama Imbroglio." In *Environmentalism in Turkey: Between Democracy and Development*, edited by Fikret Adaman and Murat Arsel, 277–90. Aldershot: Ashgate.

Arsel, Murat, Fikret Adaman, and Alfredo Saad-Filho. 2021. "Authoritarian Developmentalism: The Latest Stage of Neoliberalism?" *Geoforum* 124.

Arsel, Murat, Bengi Akbulut, and Fikret Adaman. 2015. "Environmentalism of the Malcontent: Anatomy of an Anti-Coal Power Plant Struggle in Turkey." *Journal of Peasant Studies* 42(2): 371–95.

Auyero, Javier. 2011. "Patients of the State: An Ethnographic Account of Poor People's Waiting." *Latin American Research Review* 46(1): 5–29.

Auyero, Javier, and Deborah Swistun. 2009. *Flammable: Environmental Suffering in an Argentine Shantytown*. Oxford: Oxford University Press.

Aydın, Zülküf. 2009. "De-agrarianization and De-peasantization: Dilemmas of Neo-liberal Restructuring in Turkish Agriculture." In *Turkey and the Global Economy*, edited by Ziya Öniş and Fikret Şenses, 235–54. London: Routledge.

Badem, Candan. 2018. *Çarlık Yönetimi'nde Kars, Ardahan, Artvin: 1878–1918*. Istanbul: Aras.

Baviskar, Amita, and Nandini Sundar. 2008. "Democracy versus Economic Transformation?" *Economic and Political Weekly* 43–46: 87–89.

Baxstrom, Richard. 2011. "Even Governmentality Begins as an Image: Institutional Planning in Kuala Lumpur." *Focaal* 61: 61–72.

Bayurgil, Ladin. 2019. "Caught between Political Opposition and Economic Gain: Homeowner Initiated Urban Transformation in Istanbul." Unpublished paper presented at the European Sociological Research Network, Manchester, 28 June 2019.

Bear, Laura. 2015. *Navigating Austerity: Currents of Debt along a South Asian River*. Stanford: Stanford University Press.

———. 2020. "Speculation: A Political Economy of Technologies of Imagination." *Economy and Society* 49(1): 1–15.

Bellér-Hahn, Ildiko. 2014. "The Bulldozer State: Chinese Socialist Development in Xinjiang." In *Ethnographies of the State in Central Asia: Performing Politics*, edited by Madeleine Reeves, Johan Rasanayagam, and Judith Beyer, 173–97. Bloomington: Indiana University Press.

Benjamin, Walter. 1968. *Illuminations: Essays and Reflections*. Translated by Harry Zohn. New York: Schocken Books.

———. 2002. *The Arcades Project*. Translated by Howard Eiland and Kevin McLaughlin. Cambridge, MA: Harvard University Press.

Benlisoy, Foti. 2012. "Emekçiler ve Kürtler ya da "İşçi Sınıfı Kürtleşti." Retrieved 14 August 2019 from http://fotibenlisoy.tumblr.com.

Berlant, Lauren. 2011. *Cruel Optimism*. Durham, NC: Duke University Press.

Biner, Özlem. 2020. *States of Dispossession: Violence and Materiality in Southeastern Turkey*. Philadelphia: University of Pennsylvania Press.

Biner, Özlem, and Özge Biner. 2019. *Beklerken: Zamanın Bilgisi ve Öznenin Dönüşümü*. Istanbul: Iletişim.

Bora, Tanıl. 2016. "Yerli ve Milli." *Birikim*, 6 January 2016. https://birikimdergisi.com/haftalik/7422/yerli-ve-milli.

Bratman, Eva. 2014. "Contradictions of Green Development: Human Rights and Environmental Norms in Light of Belo Monte Dam Activism." *Journal of Latin American Studies* 46(2): 261–89.

Buğra, Ayşe, and Çağlar Keyder. 2006. "The Turkish Welfare Regime in Transformation." *Journal of European Social Policy* 16(3): 211–28.

Çalışkan, Koray, and Michel Callon. 2009. "Economization, Part 1: Shifting Attention from the Economy towards Processes of Economization." *Economy and Society* 38(3): 369–98.

Campbell, Jeremy. 2014. "Speculative Accumulation: Property-Making in the Brazilian Amazon." *The Journal of Latin American and Caribbean Anthropology* 19(2): 237–59.

Casarino, Cesare, and Antonio Negri. 2008. *In Praise of the Common: A Conversation on Philosophy and Politics*. Minneapolis: University of Minnesota Press.

Çavuşoğlu, Erbatur, and Julia Strutz. 2014a. "We'll Come and Demolish Your House: The Role of Spatial (Re-)Production in the Neoliberal Hegemonic Politics of Turkey." In *Turkey Reframed: Constituting Neoliberal Hegemony*, edited by Ismet Akça, Ahmet Bekmen, and Barış Alp Özden, 141–55. London: Pluto Press.

———. 2014b. "Producing Force & Consent: Urban Transformation and Corporatism in Turkey." *City* 18(2): 134–47.

Chakravarti, Sudeep. 2017. "Narmada Dam: Public Finance, Public Play." *Mint*, 21 September 2017. https://www.livemint.com/Home-Page/NXCSsuu5Hk5QC79mn6zgYK/Narmada-dam-Public-finance-public-play.html.

Chatterjee, Partha. 2004. *The Politics of the Governed: Reflections on Popular Politics in Most of the World*. New York: Columbia University Press.

———. 2008. "Democracy and Economic Transformation in India." *Economic & Political Weekly* 43(16): 53–62.

Crehan, Kate. 2002. *Gramsci, Culture and Anthropology*. London: Pluto Press.

Davis, Diana K., and Edmund Burke III. 2011. *Environmental Imaginaries of the Middle East and North Africa*. Athens: Ohio University Press.

Demirel, Muammer. 2010. *Yusufeli Tarihi*. Istanbul: Yusufeli Municipality Press.

Dey, Ishita, and Giorgio Grappi. 2015. "Beyond Zoning: India's Corridors of 'Development' and New Frontiers of Capital." *South Atlantic Quarterly* 114(1): 153–70.

Dissard, Laurent. 2017. "From Shining Icons of Progress to Contested Infrastructures: 'Damming' the Munzur Valley in Eastern Turkey." In *Contested Spaces in Contemporary Turkey: Environmental, Urban and Secular Politics*, edited by Fatma Müge Göçek, 229–57. London: IB Taurus.

Doğan, Sevinç. 2016. *Mahalledeki AKP: Parti İşleyişi, Taban Mobilizasyonu ve Siyasal Yabancılaşma*. Istanbul: İletişim.

Doner, Fatma Nil. 2015. "Rethinking Agricultural Policy in Turkey: International Context and Domestic Developments." *Eurasian Journal of Social Sciences* 3(3): 30–51.

Eberlein, Christine. 2007. "Report on a Fact-Finding Mission to Yusufeli: On Consultation and Relocation of People Displaced by the Planned Yusufeli Dam Project on the Çoruh River in Turkey." *Banktrack*, 25 October 2007. https://www.banktrack.org/manage/ems_files/download/report_on_a_fact_finding_mission_to_yusufeli/1_071005_report_yusufeli_mission_ce_oct_07.pdf.

Eberlein, Christine, Heike Drillisch, Ercan Ayboğa, and Thomas Wenidoppler. 2010. "The Ilısu Dam in Turkey and the Role of Export Credit Agencies and NGO Networks." *Water Alternatives* 3(2): 291–312.

Eder, Mine. 2015. "Türk Usulü Buldozer Neoliberalleşmeyi Anlamak: AKP'nin Politik Ekonomisi ve Ötesi." In *Türkiye'de Yeni Iktidar Yeni Direniş: Sermaye-Ulus Devlet Karşısında Yerelötesi Müşterekler*, edited by Yahya Mete Madra, 47–56. Istanbul: Metis.

Emen, Idris. 2012. "Önce onar, sonra göm." *Radikal*, 4 November 2012.

Eren, Ceren, and Atakan Büke. 2016. "'Sen Aklini Mu Yedin Çocuk?': Neoliberalizm, Değersizleşme ve HES karşıtı Hareketler: Fındıklı Direnişi Örneği." In *Sudan Sebepler: Türkiye'de Neoliberal Su ve Enerji Politikaları ve Direnişler*, edited by Cemil Aksu, Sinan Erensü, and Erdem Evren, 313–39. Istanbul: Iletişim.

Erensü, Sinan. 2016. "Fragile Energy: Power, Nature and the Politics of Infrastructure in the 'New Turkey.'" PhD diss., University of Minnesota.

———. 2018. "The Contradictions of Turkey's Rush to Energy." *MERIP* 288: 32–35.

Erensü, Sinan, Erdem Evren, and Cemil Aksu. 2016. "Giriş: Yeğin Sular Daim Engine Akar." In *Sudan Sebepler: Türkiye'de Neoliberal Su ve Enerji Politikaları ve Direnişler*, edited by Cemil Aksu, Sinan Erensü, and Erdem Evren, 9–37. Istanbul: Iletişim.

Evren, Erdem. 2012. "The Ones Who Walk Away: Law, Sacrifice and Conscientious Objection in Turkey." In *Law against the State: Ethnographic Forays into Law's Transformations*, edited by Julia Eckert, Brian Donahoe, Christian Strümpell, and Özlem Biner, 245–65. Cambridge, UK Cambridge University Press.

———. 2016. "Bir Baraş Karşıtı Hareketin Yükselişi ve Düşüşü: Yusufeli Barajı Projesi ve Hidro-Kalkınmanın Zaman-Mekân Siyaseti." In *Sudan Sebepler: Türkiye'de Neoliberal Su ve Enerji Politikaları ve Direnişler*, edited by Cemil Aksu, Sinan Erensü, and Erdem Evren, 269–91. Istanbul: Iletişim.

Feher, Michel. 2018. *Rated Agency: Investee Politics in a Speculative Age*. New York: Zone Books.

Ferguson, James. 2005. "Seeing Like an Oil Company: Space, Security and Global Capital in Neoliberal Africa." *American Anthropologist* 107(3): 377-382.

Ferguson, James, and Akil Gupta. 2002. "Spatializing State: Towards an Ethnography of Neoliberal Governmentality." *American Ethnologist* 29: 981–1002.

Ferme, Mariane. 1998. "The Violence of Numbers: Consensus, Competition, and the Negotiation of Disputes in Sierra Leone." *Cahier d'Etudes Africaines* 150–52(2–4): 555–80.

Folch, Christine. 2019. *Hydropolitics: The Itaipu Dam, Sovereignty, and the Engineering of Modern South America*. Princeton: Princeton University Press.

Fraser, Nancy. 2014. "Behind Marx's Hidden Abode." *New Left Review* 86: 55–72.

———. 2016. "Contradictions of Capital and Care." *New Left Review* 100: 99–117.

Fujikura, Ryo, and Mikiyasu Nakayama. 2009. "Lessons Learned from the World Commission on Dams." *International Environmental Agreements: Politics, Law and Economics* 9: 173–90.

Ghosh, Kaushik. 2006. "Between Global Flows and Local Dams: Indigenousness, Locality, and the Transnational Sphere in Jharkhand, India." *Cultural Anthropology* 21(4): 510–34.

Gianighian, Raffaele. 2016. *Hodorçur: Vatanını Arayan Bir Gezginin Seyahati*. Istanbul: Iletişim.

Glastonbury, Nicholas, and Defne Kadıoğlu. 2016. '"Cleaning out the Ghettos': Urban Governance and the Remaking of Kurdistan." *Jadaliyya*, 18 March 2016. https://www.jadaliyya.com/Details/33102.

Gordillo, Gaston. 2014. *Rubble: The Afterlife of Destruction*. Durham, NC: Duke University Press

———. 2019. "The Metropolis: The Infrastructure of the Anthropocene." In *Infrastructure, Environment and Life in the Anthropocene*, edited by Kregg Hetherington, 66–94. Durham, NC: Duke University Press.

Graeber, David. 2004. *Fragments of an Anarchist Anthropology*. Chicago: Prickly Paradigm Press.

Gramsci, Antonio. 1971. *Selections from the Prison Notebooks*. London: Lawrence and Wishart.

Guyer, Jane. 2007. "Prophecy and the Near Future: Thoughts on Macroeconomic, Evangelical, and Punctuated Time." *American Ethnologist* 34(3): 409–21.

Hage, Ghassan. 2009. *Waiting*. Melbourne: Melbourne University Publishing.

Hardt, Michael, and Antonio Negri. 2009. *Commonwealth*. Cambridge, MA: Harvard University Press.

Harms, Erik. 2013. "Eviction Time in the New Saigon: Temporalities of Displacement in the Rubble of Development." *Cultural Anthropology* 28(2): 344–68.

Harootunian, Harry. 2019. *The Unspoken as Heritage: The Armenian Genocide and Its Unaccounted Lives*. Durham, NC: Duke University Press.

Harris, Leila. 2009. "States at the Limit: Tracing Contemporary State-Society Relations in the Borderlands of Southeastern Turkey." *EJTS: European Journal of Turkish Studies* 10. https://doi.org/10.4000/ejts.4122.

Harvey, David. 1982. *The Limits to Capital*. London: Verso.

———. 2003. *The New Imperialism*. Oxford: Oxford University Press.
———. 2010. *The Enigma of Capital and the Crises of Capitalism*. London: Profile Books.
———. 2011. "The Future of the Commons." *Radical History Review* 109: 101–7.

Hetherington, Kregg. 2020. *The Government of Beans: Regulating Life in the Age of Monocrops*. Durham, NC: Duke University Press.

İlhan, Akgün. 2016. "Avrupa'dan Ortadoğu'ya Ilısu Projesi'ne Karşı Hareket." In *Sudan Sebepler: Türkiye'de Neoliberal Su ve Enerji Politikaları ve Direnişler*, edited by Cemil Aksu, Sinan Erensü, and Erdem Evren. 339–57. Istanbul: Iletişim.

İslamoğlu, Huricihan. 2002. "IMF Kaynaklı Kurumsal Reformlar ve Tütün Yasası." *Birikim* 158: 20–27.

Kadirbeyoğlu, Zeynep. 2018. "Waterproof Development?: Impact of Advocacy Networks on Anti-Dam Movements in India and Turkey." In *New Perspectives on India and Turkey: Connections and Debates*, edited by Smita T. Jassal and Halil Turan, 182–95. London: Routledge.

Kalb, Don. 2017. "Frontlines: Class, Value and Social Transformation in 21st Century Capitalism." Retrieved 11 March 2019 from https://www.uib.no/sites/w3.uib.no/files/frontlines 2017.pdf.

Kalb, Don, and Gabor Halmai. 2011. *Headlines of Nation, Subtexts of Class*. New York: Berghahn Books.

Karatani, Kojin. 2014. *The Structure of World History: From Modes of Production to Modes of Exchange*. Durham, NC: Duke University Press.

Kaya, Alp Yücel. 2016. "Sermaye-Emek Kutuplaşmasının Yeniden Üretimi: Acele Kamulaştırma Kararlarında HES'ler." In *Sudan Sebepler: Türkiye'de Neoliberal Su ve Enerji Politikaları ve Direnişler*, edited by Cemil Aksu, Sinan Erensü, and Erdem Evren, 65–93. Istanbul: Iletişim.

Kaya, Ayhan. 2015. "Islamisation of Turkey under the AKP Rule: Empowering Family, Faith and Charity." *South European Society and Politics* 20(1): 47–69.

Kévorkian, Raymond. 2015. *Ermeni Soykırımı*. Istanbul: Iletişim.

Keyder, Çağlar. 2005. "Globalization and Social Exclusion in Istanbul." *International Journal of Urban and Regional Research* 29(1): 124–34.

———. 2010. "Capital City Resurgent: İstanbul since the 1980s." *New Perspectives on Turkey* 43: 177–86.

Keyder, Çağlar, and A. Öncü. 1994. "Globalization of a Third-World Metropolis: Istanbul in the 1980s." *Review* 17(3): 383–421.

Keyder, Çağlar, and Z. Yenal. 2013. *Bildiğimiz Tarımın Sonu: Küresel İktidar ve Köylülük*. Istanbul: Iletişim.

Khagram, Sanjeev. 2004. *Dams and Development: Transnational Struggles for Water and Power*. Ithaca: Cornell University Press.

Klaphake, Axel, and Waltina Scheumann. 2011. "Coruh River Basin: Hydropower Development and Transboundary Cooperation." In *Turkey's Water Policy: National Frameworks and International Cooperation*, edited by Ayşegül Kibaroğlu, Waltina Scheumann, and Annika Kramer, 251–63. Heidelberg: Springer.

Kocamaner, Hikmet. 2018. "The Politics of Family Values in Erdoğan's New Turkey." *MERIP* 288: 36–39.

Kurtiç, Ekin. 2019. "Sediment in Reservoirs: A History of Dams and Forestry in Turkey." In *Transforming Socio-Natures in Turkey: Landscapes, State and Environmental Movements*, edited by Onur Inal and Ethemcan Turhan, 90–112. London: Routledge.

Kuyucu, Tuna, and Özlem Ünsal. 2010. "'Urban Transformation' as State-Led Property Transfer: An Analysis of Two Cases of Urban Renewal in Istanbul." *Urban Studies* 47(7): 1479–99.

Laszczkowski, Mateusz. 2011. "Building the Future: Construction, Temporality, and Politics in Astana." *Focaal* 60: 77–92.

Le Guin, Ursula. 2000. "The Ones Who Walk Away From Omelas." In *The Wind's Twelve Quarters*, 275–84. London: Victor Gollancz.

Levien, Michael. 2018. *Dispossession without Development: Land Grabs in Neoliberal India*. New York: Oxford University Press.

Li, Tania Murray. 2005. "Beyond the 'State' and the Failed Schemes." *American Anthropologist* 107(3): 383–94.

———. 2007. *The Will to Improve: Governmentality, Development and the Practice of Politics*. Durham, NC: Duke University Press.

———. 2019. "Politics, Interrupted." *Anthropological Theory* 19(1): 29–53.

Lovering, John, and Hade Türkmen. 2011. "Bulldozer Neoliberalism in Istanbul: The State-Led Construction of Property Markets and the Displacement of the Urban Poor." *International Planning Studies* 16(1): 73–96.

Madra, Yahya M. 2017. "Antinomies of Globalization." *Markets, Globalization & Development Review* 2(3): 1–19.

Madra, Yahya M., and Fikret Adaman. 2014. "Neoliberal Reason and Its Forms: De-Politicization Through Economization." *Antipode* 46(3): 691–716.

Madra, Yahya M., and Ceren Özselçuk. 2019. "Capital's Sovereign Exception." *Polygraph: An International Journal of Culture & Politics* 27: 111–33.

Martinez-Alier, Joan. 2002. *The Environmentalism of the Poor: A Study of Ecological Conflicts and Valuation*. Northampton: Edward Elgar Publishing.

Marx, Karl. (1887) 1977. *Capital: A Critique of Political-Economy, Vol I*. Moscow: Progress Publishers.

Mattioli, Fabio. 2020. *Dark Finance: Illiquidity and Authoritarianism at the Margins of Europe*. Stanford: Stanford University Press.

Mezzadra, Sandro, and Brett Neilson. 2019. *The Politics of Operations: Excavating Contemporary Capitalism*. Durham, NC: Duke University Press.

Moore, Jason W. 2015. *Capitalism in the Web of Life: Ecology and the Accumulation of Capital*. London: Verso.

Morris, Rosalind. C. 2008. "Rush/Panic/Rush: Speculations on the Value of Life and Death in South Africa's Age of AIDS." *Public Culture* 20(2): 199–231.

Navaro, Yael. 2020. "The Aftermath of Mass Violence: A Negative Methodology." *Annual Review of Anthropology* 49: 161–73.

Navaro-Yashin, Yael. 2012. *The Make-Believe Space: Affective Geography in a Post-War Polity*. Durham, NC: Duke University Press.

Negri, Antonio. 2005. *Books for Burning: Between Civil War and Democracy in 1970s Italy*. London: Verso.

Nestmann, Liesa V. 1960. "Hydro-electric schemes in Turkey." *Geography* 45(3): 53–67.

Nixon, Rob. 2013. *Slow Violence and the Environmentalism of the Poor*. Boston: Harvard University Press.

Ong, Aihwa. 2000. "Graduated Sovereignty in South-East Asia." *Theory, Culture & Society* 17(4): 55–75.

Povinelli, Elizabeth. 2011. *Economies of Abandonment: Social Belonging and Endurance in Late Liberalism*. Durham, NC: Duke University Press.

Routledge, Paul. 2003. "Voices of the damned: Discursive resistance amid Erasure in the Narmada Valley, India." *Political Geography* 22: 243–70.

Sağlam, Erol. 2020. "Treasures and Discreet Afterlives of Greek Heritage in Contemporary Turkey."*Allegralaboratory.net*. Retrieved 8 October 2020 from https://allegralaboratory.net/afterlives-presumed-dead-but-lived-on/.

Sanyal, Kalyan. 2007. *Rethinking Capitalist Development: Primitive Accumulation, Governmentality & Post-Colonial Capitalism*. New Delhi: Routledge.

Scott, James. 1998. *Seeing Like a State: How Certain Schemas to Improve the Human Condition Have Failed*. New Haven: Yale University Press.

Searle, Llerena Guiu. 2016. *Landscapes of Accumulation: Real Estate and the Neoliberal Imagination in Contemporary India*. Chicago: University of Chicago Press.

Secor, Anna. 2007. "Between Longing and Despair: State, Space and Subjectivity in Turkey." *Environmental Planning D: Society and Space* 25(1): 33–52.

Şen, Besime. 2017. "Kürt İllerinde Soylulaştırma: Sur Projesi." *Basnews*. Retrieved 17 April 2018 from http://www.basnews.com/index.php/tr/opinion/konuk-yazar/363681.

Smith, Gavin. 1999. *Confronting the Present: Towards a Politically Engaged Anthropology*. Oxford: Berg.

Sneddon, Chris, and Coleen Fox. 2008. "Struggles over Dams as Struggles for Justice: The World Commission on Dams (WCD) and Anti-dam Campaigns in Thailand and Mozambique." *Society and Natural Resources* 21(7): 625–40.

Stahl, Dale. 2019. "A Technopolitical Frontier: The Keban Dam Project and Southeastern Anatolia." In *Transforming Socio-Natures in Turkey: Landscapes, State and Environmental Movements*, edited by Onur İnal and Ethemcan Turhan, 31–51. London: Routledge.

Stoler, Ann Laura, ed. 2013. *Imperial Debris: On Ruins and Ruination*. Durham, NC: Duke University Press.

Subramanian, Ajantha. 2009. *Shorelines: Space and Rights in South India*. Stanford: Stanford University Press.

Taussig, Michael. 1999. *Defacement: Public Secrecy and the Labor of the Negative*. Stanford: Stanford University Press.

Thomas, Peter. D. 2010. *The Gramscian Moment: Philosophy, Hegemony and Marxism*. Chicago: Haymarket Books.

Tomba, Massimilano. 2013. *Marx' Temporalities*. Leiden: Brill.

Tronti, Mario. 2019. *Workers and Capital*. London: Verso.

Tsing, Anna L. 2005. *Friction: An Ethnography of Global Connection*. Princeton: Princeton University Press.

———. 2015. *The Mushroom at the End of the World: On the Possibility of Life in Capitalist Ruins*. New Jersey: Princeton University Press.

Tuğal, Cihan. 2009. *Passive Revolution: Absorbing the Islamic Challenge to Capitalism*. Stanford: Stanford University Press.

Türkyılmaz, Yektan. 2011. "Rethinking Genocide: Violence and Victimhood in Eastern Anatolia: 1913–1915." PhD diss., Duke University.

Üngör, Uğur Ümit, and Mehmet Polatel. 2011. *Confiscation and Destruction: The Young Turk Seizure of Armenian Property*. London: Bloomsbury.

Verdery, Katherine. 2003. *The Vanishing Hectare: Property and Value in Postsocialist Transylvania*. Ithaca: Cornell University Press.

Von Bieberstein, Alice. 2017. "Treasure/Fetish/Gift: Hunting for 'Armenian Gold' in Post-Genocide Turkish Kurdistan." *Subjectivity* 10: 170–89.

———. 2022. "Speculating on Death: Treasure-Hunting in Present-Day Moush." In *Reverberations: Violence across Space and Time*, edited by Seda Altuğ, Zerrin Özlem Biner, Yael Navaro, and Alice von Bieberstein. Philadelphia: Pennsylvania University Press.

World Bank Report. 2018. "H1 2018: Private Participation in Infrastructure (PPI)." Retrieved 13 February 2020 from https://ppi.worldbank.org/content/dam/PPI/documents/H12018_PPI_Report.pdf.

Yaka, Özge. 2017. "A Feminist-Phenomenology of Women's Activism against Hydropower Plants in Turkey's Eastern Black Sea Region." *Gender, Place and Culture* 24(6): 869–89.

Yörük, Erdem. 2009. "Türkiye'de İşçi Sınıfı Kürtleşti." *Sendika.org*. Retrieved 19 March 2020 from https://sendika63.org/2009/12/turkiyede-isci-sinifi-kurtlesti-erdem-yoruk-ile-soylesi-ismat-kayhan-anf-38732/.

———. 2012. "Welfare Provision as Political Containment: The Politics of Social Assistance and the Kurdish Conflict in Turkey." *Politics & Society* 40(4): 517–47.

Yusufeli Yedinci Yerini Arıyor. 2017. Istanbul: Yusufeli Municipality Press.

INDEX

CPSIA information can be obtained
at www.ICGtesting.com
Printed in the USA
LVHW082055070522
718180LV00004B/131

9 781800 732803